YO-BRD-538

THE DNA DOCTOR

Candid Conversations with James D. Watson

6, 107 Hutchins www
142 ~ out of research by
1970s, 8 yrs 1948-56
1956
202 Book: Death of Life
xi Candid Science vol-1-6
www
Textbooks from 1956 on
Director Cold Springs
1968-93
Ph.D. 1950

Post 1953- RNA —then
mRNA stopped
96 our lives hn com

Also by István Hargittai

The Martians of Science: Five Physicists Who Changed the Twentieth Century. Oxford University Press, New York, 2006.

Candid Science I–VI: Conversations with Famous Scientists (with M. Hargittai and B. Hargittai) Imperial College Press, London, 2000–2006.

√ *Our Lives: Encounters of a Scientist.* Akadémiai Kiadó, Budapest, 2004.

The Road to Stockholm: Nobel Prizes, Science, and Scientists. Oxford University Press, Oxford, 2002 (soft cover 2003).

Symmetry through the Eyes of a Chemist (with M. Hargittai), Second Edition, Plenum, New York, 1995.

Symmetry: A Unifying Concept (with M. Hargittai), Shelter, Bolinas, California, 1994.

About the Author

István Hargittai is Professor of Chemistry; head of the George A. Olah PhD School at the Budapest University of Technology and Economics; and head of the Materials Structure and Modeling Research Group of the Hungarian Academy of Sciences. He is member of the Hungarian Academy of Sciences, foreign member of the Norwegian Academy of Science and Letters, and member of the Academia Europaea (London). He holds a Ph.D. from Eötvös University of Budapest, a D.Sc. from the Hungarian Academy of Sciences, and honorary doctorates from Moscow State University, the University of North Carolina, and the Russian Academy of Sciences. He has lectured in over 30 countries and taught at several universities in the United States. He has published extensively on structural chemistry and on symmetry-related topics. He and his scientist wife live in Budapest. Their grown children live in the United States.

Address: Budapest University of Technology and Economics, P.O. Box 91, H-1521 Budapest, Hungary; istvan.hargittai@gmail.com

THE DNA DOCTOR

Candid Conversations with James D. Watson

István Hargittai

Additional books
by Hargittai.
Also CSHL Press,

World Scientific

NEW JERSEY · LONDON · SINGAPORE · BEIJING · SHANGHAI · HONG KONG · TAIPEI · CHENNAI

Published by

World Scientific Publishing Co. Pte. Ltd.

5 Toh Tuck Link, Singapore 596224

USA office: 27 Warren Street, Suite 401-402, Hackensack, NJ 07601

UK office: 57 Shelton Street, Covent Garden, London WC2H 9HE

British Library Cataloguing-in-Publication Data
A catalogue record for this book is available from the British Library.

Cover design by Jimmy Low Chye Chim. Front, Portrait of James D. Watson on April 25, 2003, in Cambridge, UK (photo by M. Hargittai), and the sculpture Spirals Time – Time Spirals by Charles A. Jencks at Cold Spring Harbor Laboratory (photo by I. Hargittai); Back, James D. Watson at the Cold Spring Harbor Laboratory in June 1953 (photo by Karl Maramorosch).

THE DNA DOCTOR
Candid Conversations with James D. Watson

Copyright © 2007 by István Hargittai

All rights reserved. This book, or parts thereof, may not be reproduced in any form or by any means, electronic or mechanical, including photocopying, recording or any information storage and retrieval system now known or to be invented, without written permission from the Publisher.

For photocopying of material in this volume, please pay a copying fee through the Copyright Clearance Center, Inc., 222 Rosewood Drive, Danvers, MA 01923, USA. In this case permission to photocopy is not required from the publisher.

ISBN-13 978-981-270-797-0 (pbk)
ISBN-10 981-270-797-2 (pbk)

Typeset by Stallion Press
Email; enquiries@stallionpress.com

Printed by Mainland Press Pte Ltd

For Matthew Aaron and Stephanie Denise

Contents

Preface

In the years 2000 and 2002, I had several conversations with James (Jim) D. Watson, the Nobel laureate co-discoverer of the double helical structure of DNA. Two of these conversations were recorded on tape and excerpts from one have appeared in print.[1] There was also a third recorded conversation, conducted by my wife, Magdi. The three conversations covered a wide range of topics, including progress in science, the role of scientist in modern life, women in science, scientific ethics, terrorism, religion, and his relationship to fellow scientists. They revealed important aspects of the thinking of this major contributor to the science of our time. These conversations form the basis for this book.

Watson and Crick's "suggestion"[2] for the structure of DNA has been labeled the most important discovery in biology since Darwin and the most important discovery in science in the second half of the 20th century. Its consequences reverberate in the 21st century. Watson was also the architect of the molecular biological laboratories at Harvard University, and he built Cold Spring Harbor Laboratory into a world-class center of biomedical research. He was a principal player in the Human Genome Project, which promises to reform medicine in the decades to come. He has also been influential with his books. Watson has become a legend in his lifetime, and not only among scientists. In the neighborhood of Cold Spring Harbor, on the north shore of Long Island, he is popularly referred to as "the DNA Doctor."

Because Watson is so well known, what he thinks and says is important, and this is why I found it worthwhile to share our conversations with a broad readership. Naturally, these conversations cannot provide

---

[1] Hargittai, I. *Candid Science II: Conversations with Famous Biomedical Scientists.* Imperial College Press, London, 2002, pp. 2–15. (James D. Watson)

[2] Watson, J. D., Crick, F. H. C., "A Structure for Deoxyribose Nucleic Acid." *Nature* 1953, April 25, pp. 737–738, p. 737.

a comprehensive portrait of a scientist, especially as complex as Watson. Therefore, I augment them with comments and with excerpts from conversations with other contributors to the biological revolution, such as Erwin Chargaff, Francis Crick, Sydney Brenner, and others. These encounters have opened up for me an exciting world of modern biology. I must admit that biology bored me when I was in school although it was at the time when the double helix discovery happened and when the genetic code was being broken, but we had no idea about those advances. I became a physical chemist and have investigated the structure of small molecules. In my research, I tried to push the limits of possibilities to study small molecular structures and determine them as accurately as possible. I used to think — mistakenly, as it turned out — that the fine details of structure would not be of interest for large, biologically important molecules. In 2000, I spent three months at the MRC Laboratory of Molecular Biology in Cambridge, England, and wanted to validate this premonition, but came away with the opposite conclusion.[3]

Today, I find biomedical research to be the most exciting area of science. This change in my attitude towards the biological sciences is the strongest motivation behind creating this book. My encounters with Jim Watson, and in particular the three months my wife and I spent as his and his wife's, Elizabeth's (Liz), guests at the Cold Spring Harbor Laboratory in 2002, brought me closer to him than to any other player of the biological revolution.

Much has been written about Watson, yet our conversations with him offered something in addition to the existing literature; that something is beyond chemistry and biology, even beyond science. It is about what Watson's path has demonstrated best and what he put in this way: "go somewhere beyond your ability and come out on top."[4]

---

[3] Hargittai, M., Hargittai, I., in *Strength from Weakness: Structural Consequences of Weak Interactions in Molecules, Supermolecules, and Crystals*, eds. Domenicano, A., Hargittai, I., Kluwer Academic, Dordrecht, 2002, pp. 91–119.
[4] Watson, J. D., "Succeeding in Science: Some Rules of Thumb." *Science* 1993, 261, September 24, pp. 1812–1813, p. 1812.

Acknowledgments

This book grew out of my encounters with James D. Watson and I am grateful to Jim and Liz for their friendship and for their hospitality during our (my wife Magdi and I) three-month stay as their guests at Cold Spring Harbor Laboratory in 2002. Through informal as well as taped conversations and our correspondence, Jim gave me a unique opportunity to catch a glimpse into his thoughts. I also appreciate the opportunity of talking with other greats of modern biology in the framework of my *Candid Science** project over the years, including Sidney Altman, Seymour Benzer, Paul Berg, Sydney Brenner, Erwin Chargaff, Francis Crick, Robert Furchgott, Walter Gilbert, Aaron Klug, Arthur Kornberg, Maclyn McCarty, Matthew Meselson, Benno Müller-Hill, Marshall Nirenberg, Linus Pauling, Max Perutz, Frederick Sanger, Jens Christian Skou, Gunther Stent, Henry Taube, Charles Weissmann, Charles Yanofsky, and many others.

I appreciate assistance with information, references, other suggestions, and photographs from Anders Bárány (Stockholm University), Maureen Berejka (Cold Spring Harbor Laboratory), Ingmar Bergström (Stockholm University), the late Erwin Chargaff (New York City), Thomas Chargaff (Surry, Maine), Endre Czeizel (Budapest), Aldo Domenicano (Rome, Italy), Mikael Esmann (University of Aarhus, Denmark), Edit Ernster (Stockholm), the late Lars Ernster (Stockholm University), Annette Faux (MRC Laboratory of Molecular Biology, Cambridge, UK), Igor Gamow (University of Colorado, Boulder), Florence Greffe (Academie des Sciences de l'Institut de France), Richard Henderson (MRC Laboratory of Molecular Biology, Cambridge, UK), Nancy Hopkins (Massachusetts Institute of Technology), Tim Hunt (Cancer Research, UK), Graeme K. Hunter (University of Western

* Hargittai, I. *et al.*, *Candid Science*, Volumes I–VI. Imperial College Press, London, 2000–2006. Each volume contains at least 36 mostly in-depth interviews with famous biomedical scientists, chemists, physicists, and others.

Ontario, London, Ontario, Canada), Isabella Karle (Naval Research Laboratory), Aaron Klug (Cambridge, UK), Erik Hviid Larsen (Royal Danish Academy of Sciences, Copenhagen), Alan L. Mackay (London), Karl Maramorosch (Rutgers University, New Brunswick, New Jersey), Maclyn McCarty (Rockefeller University, New York City), Richard Marsh (California Institute of Technology, Pasadena), Marshall W. Nirenberg (National Institutes of Health, Bethesda, Maryland), the late Guy Ourisson (University of Strasbourg), Mila Pollock (Cold Spring Harbor Laboratory), Felicity Pors (Niels Bohr Archive, Copenhagen), Alexander Rich (Massachusetts Institute of Technology, Cambridge), Frederick Sanger (Swaffham Bulbeck, Cambridge, UK), Michael Sela (Weizmann Institute, Rehovot, Israel), the late David Shoenberg (Cambridge, UK), Franklin Stahl (University of Oregon, Eugene), Joan Steitz (Yale University), Gunther S. Stent (University of California, Berkeley), Tibor Szántó (Budapest), John M. Thomas (Cambridge, UK), Alex Varshavsky (California Institute of Technology, Pasadena), James D. Watson (Cold Spring Harbor Laboratory), Charles Weissmann (Scripps Florida), and Larissa Zasurskaya (Moscow State University).

At different stages, the manuscript benefited from the reviews, suggestions, and criticism (in chronological order) by Balazs Hargittai (St. Francis University, Loretto, Pennsylvania); Alex Varshavsky (California Institute of Technology, Pasadena); Charles Weissmann (Scripps Institute, Florida); Benno Müller-Hill (University of Cologne); Pál Venetiáner (Hungarian Academy of Sciences, Szeged); George Klein (Karolinska Institute); and Aaron Klug (MRC Laboratory of Molecular Biology, Cambridge, England).

It is a pleasure to acknowledge World Scientific Senior Editor Ms. Joy Quek's enthusiastic and conscientious work in bringing out this volume.

My most special thanks are due to Magdi, who continues to be the essence of my life, for her creative contribution to this book.

J. D. Watson at the Eagle Pub in Cambridge, England during the celebrations of the 50th anniversary of the discovery of the double helix, on April 25, 2003 (photo by M. Hargittai).

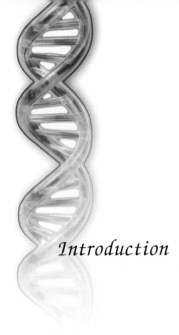

Introduction

Lucky or not, Watson was
a highly privileged young man.

Peter Medawar

It was a typical early spring afternoon at Cold Spring Harbor
Laboratory (CSHL) when a long black limousine, with James and
Elizabeth Watson in the back seats, behind its dark windows, turned
from the driveway of Ballybung house — the Watsons residence at the
northernmost tip of the CSHL campus — onto Bungtown Road, and
continued southward. It was driving slowly, majestically, by the *Time
Spirals* sculpture, a stylized version of the double helix model erected
on a mound near Ballybung. The limo passed Olney house on the
right, serving as campus security headquarters (if a visitor strayed near
Ballybung, within seconds a security car would pull up out of
nowhere, offering help). The limo continued on the semi-dirt road,
becoming a paved surface amongst laboratory and residential structures. Every building had its own history and the Watsons left their
marks on every one of them. As the limo continued further south, the
Beckman Laboratory and more conspicuously the Hazen Tower
could be seen on the right at a distance. The two together gave the
impression of a modern cathedral. What could not be seen from the

Spirals Time – Time Spirals: Double helix sculpture by Charles A. Jencks at Cold Spring Harbor Laboratory with the Watsons' residence in the background (photo by the author).

Beckman Building at Cold Spring Harbor Laboratory (CSHL) with the Hazen Tower in front of it (photos by the author).

Bungtown Road was that each side of Hazen Tower had a lowercase letter carved into it, a, c, g, and t, for the four bases of DNA, adenine, cytosine, guanine, and thymine. A bit further down on Bungtown Road, the limo passed Blackford Hall with its cafeteria where, long ago, in the late 1940s, Jim Watson used to earn his keep by serving meals to the participants of summer courses. It was also at Blackford Hall where he first reported the double helix structure of DNA at a meeting in June 1953. Finally, the limo passed Grace Auditorium dominating the scene on the right whereas on the left, at a distance, Carnegie Library could be glimpsed at, which holds the Watson archives and memorabilia. Leaving the campus, the limo turned right onto Road 25A of Long Island, merging smoothly into the light afternoon traffic, and changing gear, it sped away in the eastern direction; it was out of sight in a second. Everyone on both sides of Bungtown Road knew that Jim and Liz were heading into the City for a book launch and some also knew that they would spend some time with their older son whose chronic illness necessitated another painful hospital stay.

Whether being director, president, or chancellor, CSHL has been Watson's territory, and a long way from his humble environment in Chicago where he was born on April 6, 1928. He is most famous for his discovery, jointly with Francis Crick, of the double helix structure of DNA in 1953, in Cambridge, England. Watson and Crick, together with Maurice Wilkins, were awarded the 1962 Nobel Prize in Medicine.

The 25-year-old Watson was thrust into world fame and he has been looming over larger than life ever since. Seldom has a scientist remained at the top of science for so long after a seminal discovery and Watson's stamina is the more remarkable because he was so young when the discovery happened. As an individual, he has become emblematic of an era when the image of the lonely scientist is giving way to large and often faceless teams. Watson described the story of the discovery in his acclaimed and popular book, *The Double Helix*, to which an emblem of a previous era, William Lawrence Bragg wrote the Foreword.

Bragg was somewhat younger than Watson when he and his father founded X-ray crystal structure analysis in 1912, and he also stayed in

James D. Watson in June 1953 at Cold Spring Harbor with a model of the double helix in his left hand (photo by and courtesy of Karl Maramorosch, Scarsdale, New York).

science for a long period. When Ernest Rutherford died in 1937, Bragg was thrust into the enormously prestigious Cavendish Chair of Physics at Cambridge University. While both were physicists, Rutherford was a chemistry Nobel laureate and Bragg a physics Nobel laureate. Bragg's Cavendish leadership was hampered by World War II, and when the war ended, the Cavendish supremacy in nuclear physics — Rutherford's principal field — also ended. "Rather than fight a rearguard action,"[5] Bragg encouraged the development of such non-traditional fields of research for the Cavendish as radio-astronomy and molecular biology. Both proved to be exceptionally fruitful and were eventually signified with Nobel Prizes. Nonetheless, Bragg never attained the worldwide influence in science that Watson later acquired. Both Bragg and Watson were about 25 years old when they did their principal

---

[5] Perutz, M., *Acta Cryst.* 1970, A26, pp. 183–185, p. 185.

William Lawrence Bragg (courtesy of the late David Shoenberg, Cambridge, UK).

discoveries, but that was not unique even in the history of 20th century science. Watson himself likes to quote his one time mentor and protector, Max Delbrück, who used to emphasize that Einstein as well as Werner Heisenberg did their best science when they were 25 years old.[6] Watson has noted repeatedly that he has been called the Einstein of molecular biology. He was never too shy to appreciate himself and his achievements even though he may have appeared physically withdrawn and spoken in a barely audible voice. When he wanted to see Salvador Dali, he sent him a note saying, "The second brightest person in the world wishes to meet the brightest."[7]

---

[6] Watson, J. D., "Afterword: Five Days in Berlin." In *Murderous Science: Elimination by Scientific Selection of Jews, Gypsies, and Others in Germany, 1933–1945*, ed. Müller-Hill, B. Cold Spring Harbor Laboratory Press, New York, 1998, p. 193.

[7] Watson, *Genes, Girls and Gamow*. Oxford University Press, 2001, p. 271.

Watson was not quite a child prodigy, but had an accelerated childhood, and was catapulted into adulthood by jumping adolescence. He started school at the normal age of six; left (Catholic) religion at 12; and completed high school at 15. In 1943, he was already a student of the University of Chicago whose maverick president Robert Hutchins encouraged early entrance and sent his students back to the sources rather than feeding them with the interpretation of the classics by intermediates. Hutchins' concept of education and the intellectual fervor that characterized the University of Chicago ideally benefited Watson while others criticized this approach as hopelessly idealistic.

While an undergraduate, Watson read Erwin Schrödinger's *What Is Life?* and became hooked with the genes for life. In 1947, he began graduate school at Indiana University — an odd choice for a future revolutionary of science — because he was declined by more prestigious schools. By happy coincidence, however, Indiana University was then, for a while, a world leader in genetics. There was H. J. Muller who had just been awarded the medical Nobel Prize in 1946 "for the discovery of the production of mutations by means of X-ray radiation." Watson chose as his mentor Salvador Luria who would later become a Nobel laureate and who was a co-founder of Max Delbrück's phage school. One of Watson's fellow students was Renato Dulbecco, also a future Nobel laureate. During his graduate studies, Watson not only experienced a rapidly advancing science, but was thrust into a cosmopolitan community. Delbrück was a non-Jewish German refugee from Nazism. Muller spent three and a half years in Moscow at the famous Russian biologist N. I. Vavilov's invitation. With the rise of T. D. Lysenko, the ignorant and ruthless dictator of Soviet biology, Vavilov was to die in prison in 1943, but Muller left the Soviet Union in 1937. Luria was a Jewish-Italian refugee from Fascism; Dulbecco joined him after the war, after having studied with Professor Giuseppe Levi in Torino where Levi's other students, and in particular the future Nobel laureate Rita Levi-Montalcini had an important influence on his career.

Watson earned his PhD degree in 1950. His project was to see whether phages that had been inactivated by X-rays could be reactivated.

Jim Watson's mentors: left, Salvador Luria and his wife at CSHL in the 1950s (photo by and courtesy of Karl Maramorosch, Scarsdale, New York; note Leo Szilard on the right in the background); right, Max Delbrück (photo by and courtesy of Gunther S. Stent, Berkeley, California).

The most remarkable feature of his thesis was that he wrote it when he was only 22 years old. The lack of any exciting findings was compensated for by Delbrück telling him that this way there was no danger of falling into a trap of people wanting him to follow it up immediately. This way he would have the leisure to go on thinking and learning.[8] Watson left for postdoctoral studies in Europe where his stay in Denmark was not a great success. Then he joined the Cavendish Laboratory in Cambridge and stimulated Francis Crick to join him in their quest for the structure of DNA. Crick had been busy with writing up his PhD thesis on the structure of ox hemoglobin. There were tumultuous events preceding the discovery although they may have not appeared so significant at the time. In 1952, Watson gave a presentation — as a proxy — introducing the famous Hershey-Chase experiment, which yielded the same results as Avery *et al.*'s different (and more accurate) experiments in 1944, namely, that DNA is the substance of heredity. The scientific community was more

---

[8] Watson, J. D., "Growing Up in the Phage Group." In *Phage and the Origins of Molecular Biology*, Expanded Edition, eds. Cairns, J., Stent, G. S., Watson, J. D. Cold Spring Harbor Laboratory Press, Cold Spring Harbor, New York, 1992, pp. 239–245, p. 243.

Martha Chase and Alfred Hershey at CSHL (photo by and courtesy of Karl Maramorosch, Scarsdale, New York). The Hershey-Chase experiment yielded the same result that Avery *et al.*'s experiment in 1944, namely, that DNA was the substance of heredity, but the scientific community was better prepared to accept these findings in 1951–1952 than it was in the mid-1940s.

perceptive in 1952 than in 1944 to accept the primacy of nucleic acids over proteins, and suddenly Watson and Crick's work received more urgency. Watson later remarked that Hershey and Chase's findings "spurred me even more into finding out what DNA looked like in three dimensions."[9] It was also in 1952 that Erwin Chargaff visited the Cavendish and met with Watson and Crick. Although Chargaff was dismayed by the lack of knowledge of chemistry by the two young researchers, he shared with them his seminal findings unselfishly that the purine and pyrimidine bases were in a one-to-one ratio in the DNAs of all organisms examined to date. Watson and Crick did not recognize the structural significance of Chargaff's findings at the time — and Chargaff did not either. Eventually, Chargaff's observations became a supporting evidence for the double helix model of DNA.

---

[9] Watson, *Genes, Girls and Gamow*, p. 17.

July 1952 meeting in Royaumont, France. James D. Watson is third from the right sitting in the front row; André Lwoff is the first, sitting on the right, second row; Alfred Hershey is standing first on the right; Max Delbrück is standing 11th from the left and François Jacob is 19th; Jacques Monod is kneeling in front of Delbrück; and Seymour Benzer is on his left (courtesy of Gunther S. Stent, Berkeley, California).

Even in *The Double Helix*, Watson appears as if almost downplaying the importance of Chargaff's discoveries. In contrast, 35 years later, in *DNA: The Secret of Life*, he mentions explicitly that he "read Erwin Chargaff's paper describing his findings that the DNA bases adenine and thymine occurred in roughly equal amounts, as did the bases guanine and cytosine. Hearing of these one-to-one ratios Crick wondered whether, during DNA duplication, adenine residues might be attracted to thymine and vice versa, and whether a corresponding attraction might exist between guanine and cytosine. If so, base sequences on the 'parental' chains (e.g. ATGC) would have to be complementary to those on 'daughter' strands (yielding in this case

TACG)."[10] Chargaff's visit with Crick and Watson further stimulated their thinking of the implications of Chargaff's observations with Crick giving it more importance than Watson.

The urgency of Watson and Crick's work was enhanced by Linus Pauling's joining what Watson later described as race although Pauling never admitted it as such. However, Pauling did not lack the competitive spirit. According to Watson, Pauling did not let him use the Caltech X-ray machine on an occasion because he considered Watson a competitor.[11] In any case, Pauling published a triple-helix structure in early 1953. Although Bragg did not think it gentlemanly for the Cavendish to compete with King's College of London for solving the DNA structure, Pauling's contribution made him feel it was imperative to let all British forces mobilized. In the course of a few weeks time, Maurice Wilkins showed Watson Rosalind Franklin's excellent X-ray diffraction pattern of the so-called wet B form of DNA and Perutz showed Watson and Crick the report of King's College, which contained Franklin's data, prepared for MRC inspection. Franklin's results much facilitated Crick and Watson's work.

Once the two-chain nature of the structure, their anti-parallel arrangement — consistent with two-fold (C_2) symmetry and complementarity — and base-pairing became evident, the structure of DNA was solved and duly reported. It was not only a beautiful construction; it also suggested "a possible copying mechanism for the genetic material."[12] The initial report was followed by more detailed papers, but the essence of the discovery remained the same.

After the discovery, Watson did postdoctoral work at the California Institute of Technology; he became engaged in the study of the structure of RNA and the search for the messenger RNA. The "returns" were so much less significant than the discovery of the double helix that eventually he moved away from direct research and became an

[10] Watson, J. D. (with A. Berry), *DNA: The Secret of Life.* William Heinemann, London, 2003, p. 49.

[11] Benzer, S., "Some Early Recollections of Jim Watson." In *Inspiring Science: Jim Watson and the Age of DNA*, eds. Inglis, J. R., Sambrook, J., Witkowski, J. A. Cold Spring Harbor Laboratory Press, Cold Spring Harbor, New York, 2003, p. 17.

[12] Watson, J. D., Crick, F. H. C., "A Structure for Deoxyribose Nucleic Acid." *Nature* 1953, April 25, pp. 737–738, p. 737.

architect of science on an ever growing scale. From 1956 he had a position at Harvard University, and it was during his Harvard years that he launched his innovative textbook writing. He produced the first texts in molecular biology, starting with the *Molecular Biology of the Gene*, then, the *Molecular Biology of the Cell*, and forever changed the style of successful texts in this and related fields. His textbook was soaked in the chemistry that he so much lacked before. From 1968 to 1993 he was Director of Cold Spring Harbor Laboratory, at first parallel with his Harvard professorship, then, from 1976, full time. He was also the Director of the National Center for Human Genome Research of the National Institutes of Health between 1989 and 1992.

From time to time, Watson returned for sabbatical leaves to England, the locale of his initial and tumultuous success, and was enormously gratified when he was made into an Honorary British Knight in 2002. There were though limits to the British recognition; while former New York City Mayor Rudi Giuliani was handed the same honor directly by Queen Elizabeth II in London, Watson received it from the hands of the British Ambassador in Washington, DC. Domestic honors also came his way. Jimmy Carter awarded him the Presidential Medal of Freedom in 1977, and he received the National Medal of Science from Bill Clinton in 1997.

Fifteen years after the discovery of the double helix, Watson published *The Double Helix*,[13] which has become a classic. He stated in its Preface: "I am aware that the other participants in this story would tell parts of it in other ways, sometimes because their memory of what happened differs from mine and, perhaps in even more cases, because no two people ever see the same events in exactly the same light. ... Here I relate *my version* of how the structure of DNA was discovered."[14] (italics added) We are reminded of Leo Szilard when he contemplated putting together a history of the Manhattan Project, writing down the facts, not for having it published, but merely for

---

[13] Watson, J.D., *The Double Helix: A Personal Account of the Discovery of the Structure of DNA*. The New American Library, New York, 1969.
[14] Ibid., p. ix.

Left: Jim Watson with Peter Pauling who wrote the Foreword to *Genes, Girls, and Gamow* (by permission from Sir John M. Thomas, Cambridge, UK); right: Jim Watson autographing *Genes, Girls, and Gamow* at CSHL in 2002 (photo by M. Hargittai).

God to know about them. When a colleague noted that God might know the facts, Szilard said that this might be so, but "not *this version* of the facts."[15] (italics added)

Watson's negative portrayal of Rosalind Franklin in *The Double Helix* triggered a re-examination of Franklin's role in the 1953 discovery, and has led to its wider recognition.[16] As a result, today it is known that Watson and Crick used Franklin's experimental observations without her knowledge. She died a few years later, in 1958, without ever learning the whole truth about this incident.

Almost 35 years after the appearance of *The Double Helix*, its sequel appeared, *Genes, Girls and Gamow*.[17] In our first conversation, Watson mentions it as having already been written but not yet published. At that time he was having difficulties finding a publisher for it, which surprised me in view of his excellent record with books. It is

---

[15] *Leo Szilard: His Version of the Facts. Selected Recollections and Correspondence.* Edited by Spencer R. Weart and Gertrud Weiss Szilard. MIT Press, Cambridge, Massachusetts, and London, England, 1978, p. xvii.

[16] Klug, A., "The Discovery of the Double Helix." *J. Mol. Biol.* 2004, 335, pp. 3–26; and references therein.

[17] Watson, *Genes, Girls and Gamow*.

a detailed account of the first few years of Watson's life after the discovery of the double helix. A review in *Nature* called the book "tedious."[18] However, Watson thought that if he was to believe the comparisons of him with Einstein and the like, 100 years from now every detail of his life should be of great interest.

---

[18] Judson, H. F., "Honest Jim: The Sequel. Further Misadventures of One of the Most Influential Scientists of Our Day." *Nature* 2001, 413, October 25, pp. 775–776.

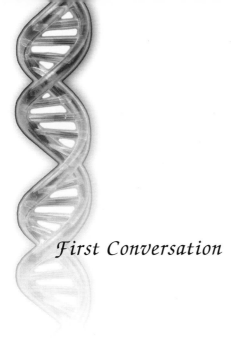

First Conversation

Major success early in life all too easily creates
the appetite for more of the same…

James D. Watson

It is of interest to record the circumstances of the interviews. I had long
wanted to interview James Watson in the framework of my project of
interviews with famous scientists.[19] When I first wrote him, he politely
but firmly declined; in his letter of February 8, 2000, he said that he
would not like to be interviewed until the publication of the first of his
two autobiographical books he was just finishing. He estimated that the
book would be out 15 months from now. This did not seem very promising. Soon, however, a second letter from him, dated April 25, 2000,
arrived in Budapest, suggesting a meeting at a certain hour on a certain
day in London. I was surprised not only by his willingness for the interview but also by his assumption that I could just take off for London for
a one-hour meeting. I was not a professional interviewer. I interviewed
people when I attended scientific meetings or when we went for family
vacations, and the interviews were side products of the occasion. So I
wrote back that I would not be in London, but — as it happened —
would be in the New York area soon.

---

[19] Hargittai, *Candid Science I–VI* book series.

James D. Watson in his office at CSHL in 2000 during the first interview (photo by
M. Hargittai).

Magdi and I visited Watson on a Saturday, on May 20, 2000, in
the Cold Spring Harbor Laboratory (CSHL). The date was selected
to fit his busy schedule, and it was the last day of our visit to the
United States. We took the train from Manhattan to Syoset Station
from where we took a cab to CSHL. There was a light rain but the
scenery was beautiful. We arrived at the campus well ahead of time,
and walked around. We found the Beckman building where Watson's
office was supposed to be. Most doors were closed but one was open,
and a foreign postdoctoral fellow ("postdoc") looked at us with sur-
prise when we told him that we were looking for Dr. Watson. He
found it hard to believe that we were going to see him because — as
he put it — Watson's presence was constantly felt but rarely experienced by him. Watson's office was closed and we waited. The time of

the appointment came, but there was no Watson, and precious minutes went by. It was not only that Watson's secretary stipulated one hour for our meeting, it was also that our departure for Budapest later on the same day restricted our flexibility.

Finally, Watson arrived; he was only about 15 minutes late, but it was one quarter of my allotted time. The interview started very slowly. Later on he told me that he had felt reluctant to give this interview as he is usually reluctant to give interviews, but two of my former interviewees, Benno Müller-Hill and Walter Gratzer had told him that he should see me. Nobody ever tells Watson what he has to do, so it was my luck that he consented. Müller-Hill was a postdoc in his lab in the 1960s and then went on for a distinguished career as Professor of Genetics at the University of Cologne. Watson wrote a substantial "Afterword" for the English edition of Müller-Hill's book *Murderous Science*.[20] Gratzer is Emeritus Professor of Biophysics at King's College, London University; he edited Watson's recent book, *A Passion for DNA*.[21]

In my interviews I usually asked the interviewee about his or her award-winning research. In the Watson interview, I did not want to waste time on this in view of the well known story of the double helix discovery. Often, the interviews take off on their own after some initial questions. In this case, however, I could not count on Watson's talkativeness. I knew he would be sparse with words because Müller-Hill had warned me about it. Müller-Hill described the atmosphere in Watson's lab at Harvard University in the 1960s. In the quoted paragraph below, "Wally" is Walter Gilbert, a theoretical physicist turned molecular biologist who started his career in molecular biology in his friend Watson's laboratory. This is what Müller-Hill said,[22]

"...there was Wally with whom I could discuss the experimental details, and he loved to talk things over. On the other hand,

---

[20] Watson, J. D., "Afterword: Five Days in Berlin." In *Murderous Science: Elimination by Scientific Selection of Jews, Gypsies, and Others in Germany, 1933–1945*, ed. Müller-Hill, B. Cold Spring Harbor Laboratory Press, New York, 1998, pp. 185–200.
[21] Watson, J. D., *A Passion for DNA: Genes, Genomes, and Society.* Oxford University Press, 2000.
[22] Hargittai, *Candid Science II*, p. 128 (Benno Müller-Hill).

there was Jim Watson, who did not talk at all. When he came into the lab, you would say one or two sentences, and that was it. You just realized that there was nothing more to say. For example, once I got a particular mutant and I was very proud of that but you can explain such an experiment in two minutes. There was another German postdoc there whom I knew from Freiburg, Klaus Weber. We had a bet about who spoke how much with Jim. After half a year, he was at 22 minutes and I was at 17 minutes, total. Jim Watson's non-speaking was even more driving. It had the effect that there was someone with whom you could speak only if you had something to say. If you had no result, there was nothing to speak about."

Fortunately, I was prepared for the Watson interview, even better than I usually was for my interviews. Reading *A Passion for DNA*, I scribbled down many questions although I did not anticipate that there would be time for all, but there was. It was not that the time was so long, rather, it was that the answers were often very short and offered no continuation.

However, at some point during the interview, something changed. The strain in the atmosphere disappeared, and the interview became a conversation. Re-reading the transcripts it is difficult to pinpoint the turning point and it might have been more gradual than my memory would suggest. In any case, the conversation did not stop at the top of the hour although there were already people waiting for Watson. When we finished, Watson drove us to the train station, and made us promise to return for what he called a more substantial visit. That he meant it became clear soon enough. But first, here is how the interview started:

What turned you originally to science?

I was just curious of why the world was like what it was? Laws of nature. Why did things happen?

Any particular book or teacher?

There was a book about bird migration, which I got for Christmas when I was seven years old. It made me interested in birds. My father had been a bird migration watcher for years. Then, when I got a little older, the question "What is life?" always seemed to be paramount. That, combined with my father's fairly strong anti-religious views provided me the perfect background. I never had to rebel against my parents; I never had any crises over beliefs.

Have you been open about being non-religious?
Yes.

Isn't this rather difficult in the United States?
Not in the Richard Dawkins sense. When I was a child I was no different from [Richard] Feynman. You don't accept Christ by revelation.

Watson read *Arrowsmith* by Sinclair Lewis early on.[23] It is about Martin Arrowsmith, a tough young man hell-bent on becoming a microbe hunter. Watson did not know at the time that Paul de Kruif helped the celebrated author writing the science-related parts. This apprenticeship to Sinclair Lewis was a great opportunity for de Kruif,[24] who was on his way from a PhD research scientist to becoming one of the most successful science writers of all time. Paul de Kruif started writing his *Microbe Hunters* in 1923; it appeared in 1926 and has been in print ever since.[25] Many future outstanding chemists, biochemists, and biomedical scientists of the second half of the 20th century turned to science in their childhood after having read de Kruif's book.

The question "What is life?" is, of course, a reference to Schrödinger's influential book *What Is Life?*[26] To me, it was especially interesting how Kary Mullis,[27] the discoverer of PCR, the Polymerase Chain Reaction (used for amplifying DNA pieces easily), referred to

---

[23] Lewis, S., *Arrowsmith*. Harcourt, New York, 1925.
[24] De Kruif, P., *The Sweeping Wind: A Memoir*. Harcourt, Brace & World, New York, 1962.
[25] De Kruif, P., *Microbe Hunters*. Harcourt Brace, New York, 1926.
[26] Schrödinger, E., *What Is Life? The Physical Aspect of the Living Cell*. Cambridge University Press, 1944.
[27] Hargittai, *Candid Science II*, pp. 182–195 (Kary B. Mullis).

this book when he wanted to explain that he did not grow up in an intellectually inspiring environment. As he was growing up in a home in Columbia, South Carolina, nobody noticed that they did not have a copy of *What Is Life?* Mullis was born in 1944, the year when the book appeared. Watson and Mullis both came from humble backgrounds, but Watson's family environment provided the intellectual stimulus Mullis may have missed. Their lives took different turns while retaining similarities as well. Mullis always dreamed of meeting Feynman and later Crick, and he did not live too far from either, but never decided to visit them until it was too late. Watson had a lot of interactions with both. Watson and Mullis are both apt to make shocking statements to their audiences, but the similarity ends here. Mullis's statements are sometimes anti-science and often self-destructing. In contrast, however shocking he may appear at times, Watson knows exactly what he can say *publicly* — like fat women have better sex lives — and what he cannot (about politics and religion, for example).

Coming back to Schrödinger's book, it was so influential because it succinctly posed a most important problem, and also because the person who posed the problem was a world renowned physicist, a veteran Nobel laureate. Others had wondered on the nature of hereditary material and came even close to declaring it to be nucleic acids on an intuitive basis. As early as 1926, J. B. Leathes speculated about the role of nucleic acids in a paper in *Science*:[28]

> "But in the chemical make-up of protoplasm, proteins, the most abundant component, are not the only ones that are necessary. Pre-eminent among others are the nucleic acids. When we consider what has been learned of the behavior and of the chemical composition of the nuclear chromosomes, and that according to Steudel's reckoning the nucleic acids form 40% of the solid components of these chromosomes, into which are packed from the beginning all that preordains, if not our fate and fortunes, at least our bodily characteristics down to the color of our eyelashes, it becomes a question

---

[28] Leathes, J. B., "Function and Design." *Science* 1926, 64, pp. 387–394.

whether the viruses of nucleic acids may not rival those of amino acid chains in their vital importance."

Many years later, J. D. Bernal in his review of Watson's *The Double Helix* would go back as far as Lucretius, 2000 years ago, to pinpoint in his *De rerum natura* something that in modern terms would be the unalterable genes where Lucretius speaks about unalterable atoms:[29]

> "No species is ever changed, but remains so much of itself that every kind of bird displays on its body its own specific markings. This is a further proof that their bodies are composed of changeless matter. For, if the atoms could yield in any way to change, there would be no certainty as to what could arise and what could not, ... nor could successive generations so regularly repeat the nature, behavior, habits and movement of their parents."

The question about being openly non-religious came up early in our conversation. Watson's categorical "Yes" to my question about being openly non-religious seems surprising in view of his usual restraint concerning religion in public statements. Generally, he has been very careful about this topic being the principal fund-raiser of Cold Spring Harbor Laboratory. However, he gave also unambiguous responses for an Italian publication in October 2002.[30] When asked about ethical questions in biology, referring to stem cells and cloning, Watson stressed that for the non-religious, these represented no problems and added that he was non-religious. He did not think in terms of offending natural laws that were a product of evolution. He considered himself lucky being without God, so he did not have to think about such things. For him the only problem was whether we wanted

---

[29] Bernal, J. D., "The Material Theory of Life." *Labour Monthly* 1968, pp. 323–326, p. 323.

[30] Odifreddi, P., *Incontri con Menti Straordinarie* (In Italian, Encounters with Extraordinary Minds). Longanesi, Milano, 2006, "James Watson," pp. 166–171, p. 171. Odifreddi did not prepare the transcripts of the conversation in English, but went directly to an Italian-language presentation of the interview (private e-mail communication from Odifreddi, P., June 2006). I thank Aldo Domenicano, Rome, for the translation of the Italian text into English.

to improve the quality of life without doing any evil to those who were close to us. He never liked the alliance of the Roman Catholic Church with Fascism, and neither did he like the Pope, Watson added. When asked specifically about John Paul II, to whom the interviewer referred as having made an opening to science, Watson told him that all the popes had the same great confusion in their heads.

Now we continue with our interview:

What would capture a child's attention today to science?

The brain. How does it work? Why do you like the taste of chocolate? There're so many problems, which are totally unexplainable. What does taste mean? What is consciousness? How do you encode a system by which you like something or you don't like it? I've started thinking about this more than superficially and I realize it's a very difficult problem.

How do you feel about classical training versus going right to the frontier? Following the double helix, Lord Todd congratulated you as an organic chemist when you were certainly not deep in organic chemistry.

He realized that our discovery was a chemistry discovery; it was not a discovery in biology.

Vladimir Prelog [see, below], not long before he died, ascribed the chemists' persistent staying away from DNA to the fact that chemists used to consider these systems to be dirty mixtures. So Lord Todd appears to have been exceptional.

Of course, he'd done a lot of chemistry of DNA. In retrospect you may ask why Todd wasn't curious about what DNA would look like in three dimensions. At that time big molecules were for "colloid chemists." They were too big to study by conventional techniques of the organic chemist. They were a different world.

The organic chemist, Alexander Todd's role in the discovery of biological macromolecules is in itself interesting. In 1951, he and his student,

Donald M. Brown in Cambridge, England, 2003 (photo by the author). Brown and A. R. Todd established in 1952 that the phosphate group is linked differently to the sugars on either side — thus the polynucleotide chain has directionality.

Donald M. Brown established the chemical structure of DNA. Todd was awarded the Nobel Prize in 1957. Todd worked near Max Perutz's laboratory in Cambridge and Perutz consulted him in chemistry when working on the structure of hemoglobin. Perutz had learned from Todd that the peptide bond was to a great extent a double bond and that should have led Perutz to suppose a planar arrangement of the bonds about the peptide bond. Alas, Perutz did not realize the relationship, and he and Bragg and Kendrew did not report the right structure for alpha-keratin, although they communicated quite a few models for it.[31] On the other hand, Linus Pauling knew the relationship between double bonds and planarity, and he came up with only one model, but it was the right one.[32]

---

[31] Bragg, W. L., Kendrew, J. C., Perutz, M. F., "Polypeptide Chain Configuration in Crystalline Proteins." *Proc. R. Soc.* 1950, 203A, pp. 321–357.

[32] Pauling, L., Corey, R. B., Branson, H. R., "The Structure of Proteins: Two Hydrogen-Bonded Helical Configurations of the Polypeptide Chain." *Proc. Natl. Acad. Sci. USA* 1951, 37, pp. 205–211.

To understand that the bonds take a planar arrangement about a double bond, the so-called theory of resonance may be invoked. This theory explains, for example, the structure of the benzene molecule, which could be described by two extreme forms in which double bonds and single bonds alternate in the ring. The two forms are equivalent, but in reality only one form exists and it can be envisioned as a resonance between the two forms. Pauling was one of the pioneers of this theory and this exposed him to much criticism in the Soviet Union where the theory of resonance was considered a manifestation of idealistic ideology in the early 1950s. It may seem surprising that an innocent theory in chemistry would generate a violent political reaction. In the background, however, there was the fear of Soviet officialdom that Western science would be making headways in Soviet science. Chemists applying this theory were condemned to be lackeys of bourgeois influence and some lost their jobs as a consequence. In other fields, notably in biology, accusations of getting under Western influence led to more tragic consequences under Stalin. Ironically, Linus Pauling was politically on the left, but when

Left: Ava and Linus Pauling (photo by and courtesy of Karl Maramorosch, Scarsdale, New York); right: Linus Pauling lecturing in Moscow in 1984 (photo by and courtesy of Larissa Zasurskaya, Moscow).

the struggle against the theory of resonance culminated in 1951, this was not known in the Soviet Union.[33]

As to the question whether the discovery of the double helix structure of DNA was a chemical discovery, initially it was, although most chemists did not consider nucleic acids to be part of chemistry. To them these macromolecules of uncertain composition were a mess on which they did not want to waste their clean techniques. As a consequence, the double helix and the ensuing scientific revolution became part of biology rather than chemistry. Of course, it does not matter in principle, but it does matter in practice, in particular in getting good students and sufficient funding for research.

Albert Eschenmoser, one of the foremost organic chemists of the late 20th century at the Swiss Federal Institute of Technology, asked Vladimir Prelog, the emblematic figure of natural products chemistry, the question "Why did you ignore DNA?" adding that "Every year during which we did not work on DNA was a wasted year."[34] For years, Eschenmoser prodded Prelog for a response until one day, at the end of his long life, Prelog handed Eschenmoser a written statement about it:[35]

"For some time you have prodded me to tell you, why the great Leopold [Ruzicka] and I did not recognize, in a timely fashion, that the nucleic acids are the most important natural products, and why did we waste our time on such worthless substances as the polyterpenes, steroids, alkaloids, etc.

My lightheaded answer was that we considered the nucleic acids as dirty mixtures that we could not and should not investigate with our techniques. Further developments were, at least in part, to justify us.

As a matter of fact, for personal and pragmatic reasons, we never considered working on nucleic acids."

[33] See, for example, Hargittai, I., *Candid Science: Conversations with Famous Chemists.* Imperial College Press, London, 2000, pp. 8–13 ("The Great Soviet Resonance Controversy").

[34] Hargittai, I., *Candid Science III: More Conversations with Famous Chemists.* Imperial College Press, London, 2003, p. 102. (Albert Eschenmoser)

[35] Ibid., p. 107.

To the question about who operates with dirty materials and clean techniques, here is a comment by the Soviet-Russian physical chemist and Nobel laureate Nikolai Semenov, according to whom what they "used to say about the difference between physics and chemistry that physicists deal with dirty materials but employ clean techniques, whereas chemists deal with clean materials but employ dirty techniques."[36] Arthur Kornberg's remark about the difference between chemists and biologists well augments the above discussion: "Chemists seek precise answers to well-defined problems, whereas biologists are content with approximate answers to complex problems."[37]

In the Watson conversation, we soon turned to the genetic code:

In your book A Passion for DNA, *Walter Gratzer says, "hunt for the genetic code was the most exciting period in early molecular biology."*

Yes, it was.

Was it really, or was it because you and a few other very influential people found it very puzzling?

No, no, it was. We wanted to know how cells read the information in DNA and you had to know the nature of the information and what the cellular machinery was. It turned out to be first RNA polymerase and then the whole ribosome.

Do you recall what happened with Marshall Nirenberg's paper in Moscow in 1961? Did people indeed rush out of the room, after its second presentation, to get home fast and repeat and continue Nirenberg's experiment?

No, I think the only person who wanted to repeat Nirenberg's experiment was Wally Gilbert.

---

[36] Hargittai, *Candid Science*, p. 472. (Nikolai N. Semenov)

[37] Kornberg, A., *The Golden Helix: Inside the Biotech Ventures*. University Science Books, Sausalito, California, 1995, p. 14.

Not Severo Ochoa?

Probably Ochoa too, because he'd made the synthetic co-polymers. But I don't recall anyone except Wally Gilbert who changed what he was doing.

You don't mention Nirenberg in your book.

I mention him in the successor to *The Double Helix.* It's been written but not published.[38] Nirenberg was very important and deserved his Nobel Prize but he was never a friend or anything like that, and he was a biochemist whereas we always thought in terms of information. That was slightly different.

Francis Crick told us on the occasion of our visit with the Cricks, on February 7, 2004, that one of his most important findings had never been written up and was recorded only in a manuscript for a lecture which seems to have been lost. This was about the one-dimensional sequence of the amino acids in proteins and the importance of the three-dimensional structures of the proteins. The sequence, which is one-dimensional, determines the folding, which produces the three-dimensional structure. In terms of replication, the three-dimensional structure could not replicate itself; the only part that is capable of replication is the *surface* of the three-dimensional structure. The essence of Crick's idea was that for replication it sufficed to repeat the sequence. Crick told us about this when I raised the question about who brought up for the first time the idea that the nucleic acids code for the proteins.

When we talked about the connection between nucleic acids and proteins, Crick said that Watson and he were definitely thinking about that in the spring of 1953. When they announced in the Cambridge pub, The Eagle, that they solved the secret of life, they could make such an announcement only because they realized that there was a connection. The double helix structure of DNA by itself would not have sufficed for calling it the secret of life. They understood the implication of the double helix so quickly because they had thought about *information*

---

[38] This was to be Watson, *Genes, Girls and Gamow.*

transfer. Nine years later, this expression found its way into the announcement of the Nobel Prize for Crick, Watson, and Wilkins.

Actually — Crick told us — he had had this idea even before he had met Watson. This is fascinating as we may try to delineate their contributions to their joint discovery. When they worked together, they talked a lot to each other, so it is hardly possible to delineate their contributions in the process of the discovery. However, the idea about the importance of sequence in replication, whether it is the sequence in nucleic acids or proteins, seems to have occurred to Crick independent of Watson. This is though not to say that others had not come to similar ideas. Even before Crick and Watson's work on the DNA structure had commenced, Erwin Chargaff wrote in 1950 — referring to his 1947 lecture in Cold Spring Harbor[39] — that "minute changes in the nucleic acid, e.g. the disappearance of one guanine molecule out of a hundred, could produce far-reaching changes in the geometry of the conjugated nucleoprotein; and it is not impossible that rearrangements of this type are among the causes of the occurrence of mutations."[40]

We also talked about Watson and Crick enumerating the 20 naturally occurring amino acids. Sometimes it happens that important findings do not appear so important at the time of their being made. Today, this enumeration is there in every textbook, but seldom is it associated with Crick and Watson or any other name for that matter.

Following the discovery of the double helix structure and its genetic implication, the question about information transfer from DNA to the proteins, called the genetic code, came up, and kept Watson's world in excitement for years. There were theories and experiments in the leading chemistry, biology, and biochemistry laboratories of the world, but the first step in breaking the genetic code was made by Marshall Nirenberg, a young and little known investigator and his German postdoc, Heinrich Matthaei, at the National Institutes of Health. Nirenberg was an unlikely discoverer in the eyes of many of the leading scientists and could not even get his report

[39] Chargaff, E., in "Nucleic Acids and Nucleoproteins." *Cold Spring Harbor Symp. Quant. Biol.* 1947, 12, p. 28.
[40] Chargaff, E., "Chemical Specificity of Nucleic Acids and Mechanism of their Enzymatic Degradation." *Experientia* 1950, 6, pp. 201–209.

Marshall W. Nirenberg in the 1960s (courtesy of Marshall W. Nirenberg, Bethesda, Maryland) and at the National Institutes of Health (NIH) in 1999 (photo by the author).

accepted for presentation at a Cold Spring Harbor meeting in 1961. Even Leo Szilard, the nuclear physicist temporarily turned biologist, who was so open to fresh ideas, did not recognize the importance of Nirenberg's discovery (see later).[41] However, when Nirenberg applied to the organizers of the Fifth International Congress of Biochemistry convened for September of the same year in Moscow, he was accepted to attend and contribute a paper. The Moscow meeting did not promise any scientific breakthrough to report, but the Western participants went there anyway because they wanted to have a glimpse of communism in action. Here are Watson's words in *Genes, Girls and Gamow* describing Nirenberg's bombshell:[42]

> "Soon, however, I heard rumors that there might be an unexpected bombshell talk by Marshall Nirenberg from the National

---

[41] Hargittai, *Candid Science II*, p. 140. (Marshall W. Nirenberg)
[42] Watson, *Genes, Girls and Gamow*, p. 265.

Institutes of Health. It was not one of the main presentations, and only a few individuals, including Alfred Tissières and Wally Gilbert, were alerted about it by its title 'The dependence of cell-free protein synthesis in *E. coli* upon naturally occurring and synthetic template RNA.' Using Alfred's improved recipe for cell-free protein synthesis, Nirenberg and his German colleague Heinrich Matthaei had found over the past several months that addition of poly(U) promoted the synthesis of polypeptides made up exclusively of the amino acid phenylalanine.

When François Jacob heard about the experiments from me over breakfast the next morning, he thought I was perpetuating a practical joke. But Nirenberg and Matthaei had done their experiments well, and Francis hastily arranged a big lecture on the last congress day that let Nirenberg convince as well as stun most in the audience. From that moment on, it seemed likely that the genetic code would soon be cracked through observing the polypeptides made in cell-free systems programmed with appropriate synthetic polyribonucleotides."

I have not found yet two identical descriptions of what happened in Moscow. According to legends, some leading scientists rushed directly from the lecture hall to the airport to return to their laboratories to repeat the experiments. This should not be taken literally; in the Soviet

Jim Watson (on the left) and Francis Crick (on the right) with Russian molecular biologist Jacob [then Yakov] Varshavsky at the Fifth International Congress of Biochemistry, Moscow, 1961 (photos courtesy of Alex Varshavsky, Pasadena, California).

Union at that time it was not a trivial matter to change one's reservations. But according to reliable sources, Nirenberg's communication electrified the atmosphere. Charles Weissmann heard Francis Crick use the word "thunderstruck" to describe his reaction to Nirenberg's report.[43] Nirenberg won a share of the 1968 Nobel Prize in Medicine.

Here, I am describing the event based on my conversation with Marshall Nirenberg in 1999:[44] In the late 1950s, protein synthesis was one of the hottest topics in biochemistry. The best biochemists in the world were working on the biosynthesis of proteins. Transfer RNA was just discovered and so were the amino-acid-activating enzymes that catalyzed the activation of transfer RNA to link an amino acid to a particular species of transfer RNA. It was known that proteins were synthesized on ribosome particles in the cells, but nobody knew anything about the messenger. Things were moving rapidly and Nirenberg thought that protein synthesis would be solved within two years though he did not expect that he would play a crucial role in it. He asked himself, "What chance do I have as a single person against the best people with big groups in the best laboratories of the world who were working on protein synthesis?"

What Nirenberg and Matthaei found at the National Institutes of Health was that viral RNA was a tremendous stimulator of amino acid incorporation into a protein. Heinz Fraenkel-Conrat at Berkeley, a great authority on tobacco mosaic virus, had some mutants that Nirenberg thought were directing the synthesis of viral coat proteins. He went to Berkeley to work with Fraenkel-Conrat for a month, using viral RNA to direct protein synthesis. He packed a bag of enzyme-containing extracts in a picnic basket and brought it on the plane. This was at the beginning of 1961. Before he left, he wrote a series of protocols for Matthaei, using synthetic polynucleotides in a cell-free system, and Matthaei was actually the person — in Nirenberg's absence — who did the first poly(U) experiment, producing polyphenylalanine. When this happened, he called Nirenberg, who returned to Bethesda at once.

---

[43] Hargittai, *Candid Science II*, p. 487. (Charles Weissmann)
[44] Ibid., pp. 136–138 (Marshall W. Nirenberg).

When initially Nirenberg and Matthaei made the first step in breaking the code, Nirenberg knew that in order to have people believe in such an experiment, of course, a precipitate had to be produced and its properties had to be studied. He knew that he might meet with disbelief; it was such a staggering discovery. Nirenberg knew that he had to "prove it upside, downside, and every which way."[45] He did not know anything about the physical properties of polyphenylalanine, and he wanted to find them and demonstrate them. They isolated the product and produced the precipitate in trichloroacetic acid.

However, Nirenberg wanted to have more and decided to go to the library to find out what was known about polyphenylalanine. On his way to the library he went on the floor beneath his lab where Christian B. Anfinsen had his laboratory. Anfinsen later (1972) became a Nobel laureate, but he was not part of Nirenberg's story. He was not even there, but a young visitor, Michael Sela, was; he later became the president of the Weizmann Institute in Israel. Nirenberg knew that Sela was knowledgeable about synthetic peptides, which he used in immunology. Nirenberg asked him whether he knew anything about the physical properties of polyphenylalanine? Sela told him that polyphenylalanine was very insoluble in normal solvents, but it dissolved in a solvent that consisted of 15% hydrobromic acid dissolved in glacial acetic acid. Sela even had some of this solvent and Nirenberg tested his precipitate in this solvent, which was completely unknown to him. At that time it did not occur to him to ask Sela how he happened to know about such a strange solvent. Fifteen or 20 years later, he ran into him and found out that Sela was probably the only person aware of that solvent at the time, which had been once prepared by mistake,[46] and then observed that it dissolved polyphenylalanine. Nirenberg wondered in our 1999 conversation about the probability that he would ask the only person in the world who could answer his question.

---

[45] Conversation with Marshall W. Nirenberg, unpublished records.

[46] Sela, M., "My World Through Science." In *Selected Topics in the History of Biochemistry: Personal Recollections VIII* (*Comprehensive Biochemistry* Vol. 43), eds. Semenza, G., Turner A.J. Elsevier, 2004, pp. 1–100, pp. 15–16.

Left: Soviet stamp commemorating the Fifth International Biochemistry Congress in Moscow, 1961; right: British stamp about cracking the genetic code.

As Nirenberg was preparing for the Moscow meeting in 1961, he knew that he had a terrific thing to report. However, being unknown in the field, he was only assigned ten minutes in a tiny room with a Soviet-style giant-size projector, and there was only a handful of people attending. Nirenberg had introduced himself to Watson before the talk and told him about the discovery. Watson sent a colleague, Alfred Tissières, to Nirenberg's talk, and he corroborated what Nirenberg had told Watson.

Moscow was a most unlikely venue in 1961 where to have a groundbreaking announcement in molecular biology. In the Soviet Union, Lysenko still dominated the scene of biological sciences though he no longer decided life and death questions, literally, as he did under Stalin. This was Khrushchev's time of thaw, but Lysenko's unscientific views still hindered progress. He denied the existence of gene — saying that it could not have existed because nobody has ever seen one. Ironically, he was director of the Institute of Genetics in Moscow. It was a sign of change though that the world congress of biochemists could take place in Moscow at all. The Western organizers were assured that they would have freedom in choosing speakers and topics though not the titles of the sections — the term "molecular biology" was still illegal in Soviet science. Vladimir Engelhardt

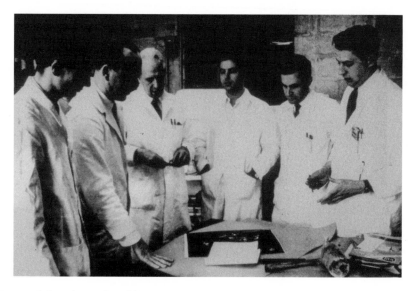

Severo Ochoa (center) and his associates in the 1960s at New York University (courtesy of Charles Weissmann, Jupiter, Florida).

remembered two decades later that they intended to name one of the sections "molecular biology," but he could not get it through and a compromise name was chosen, "biological functions at the molecular level."[47] The bureaucrats in Moscow still had the upper hand over the scientists, probably fearing that too much Western influence might creep into their sterile Soviet order.

Watson, Crick, Perutz, Severo Ochoa, other luminaries, and many lesser known scientists attended the Moscow meeting. Ochoa was one of the world's leading biochemists who won half of the 1959 medical Nobel Prize — with Arthur Kornberg the other half — "for their discoveries of the mechanisms in the biological synthesis of ribonucleic acid and deoxyribonucleic acid," according to the announcement of the Karolinska Institute. As it turned out, eventually, Ochoa did not get it right and it took several more years before the mechanism of RNA synthesis was understood in the works of others. However, nobody ever questioned that Ochoa deserved the highest recognition

---

[47] Engelhardt, W. A., "Life and Science." *Annu. Rev. Biochem.* 1982, 51, 1–19.

Jim Watson, Arthur Kornberg (in the middle), and the author in Stockholm, 2001 (photo by M. Hargittai).

for his achievements in biochemistry. Ochoa's work that the award committee singled out for recognition was a joint effort with Marianne Grunberg-Manago, but she was not included in the prize. As it happened, the enzyme Kornberg identified as responsible for DNA replication eventually turned out not to have this function either, but was responsible for DNA repair. Again, the science community has felt that Kornberg deserved the recognition in spite of this misinterpretation of his original results. Both Kornberg and Ochoa displayed Watson-Crick-style double-helix structures of nucleic acids in their respective Nobel lectures.

Nirenberg's discovery was only the first step in cracking the genetic code. After the Moscow congress, a race developed about who could correlate amino acids with triplets of bases faster. A strong group of Ochoa and his laboratory at New York University played a prominent role in this race, but there were sad side stories in it as well.[48] One of Ochoa's associates, a young Hungarian refugee scientist, later Yale professor, Peter Lengyel had wanted to carry out experiments

---

[48] Hargittai, *Candid Science II*, pp. 484–486.

similar to Nirenberg's at about the same time or even before, but let his colleagues talk him out of it. There had been though some attempts experimentally in the same group. Mirko Beljanski, a French molecular biologist had taken a sabbatical year in Ochoa's laboratory. Ochoa had asked him to experiment with polyadenylic acid, poly A, as a template for protein synthesis in cell-free extracts of *E. coli*. This was independent of Nirenberg though it was a similar experiment, and it had actually happened before the successful Nirenberg-Matthaei experiment. Beljanski spent a year on this problem with no success. It is known today that poly A codes for the protein poly-lysine, and Beljanski may have well produced it, but as it did not pre-cipitate in trichloroacetic acid — the universal solvent people used to precipitate proteins — it may have been there without Beljanski's knowing about it. So Ochoa and Beljanski missed the discovery. But Ochoa continued as a competitor after Nirenberg's Moscow announcement. Given his fame and authority, let alone his Nobel Prize, there were people who felt that he should have let the little known Nirenberg harvest the fruits of his discovery.

Nirenberg never became a celebrity or a household name outside a narrow circle of scientists. He figures though in Horace F. Judson's monumental book *The Eight Day of Creation*. Characteristically, the title of the relevant chapter is "He wasn't a member of the club."[49] Incidentally, Nirenberg finds it an "awful book. I read the parts on coding and it was not the way he said it was. His description of how the code was deciphered is terrible. He came and talked with me and I told him how it was and he wrote it in a different way. He came sev-eral times. It's a terrible account. He didn't catch the flavor in the lab. It took about six years to decipher the genetic code, between 1961 and 1966. We worked flat out, as hard as we could on this. I worked with about 20 postdoctoral fellows, no more than nine at once. They all participated in deciphering the genetic code. It was a group project. The atmosphere in the lab was most exhilarating. Everything we touched, pretty much worked. Discoveries were coming fast and furious.

---

[49] Judson, H. F., *The Eighth Day of Creation: Makers of the Revolution in Biology*, Expanded Edition. Cold Spring Harbor Laboratory Press, Cold Spring Harbor, New York, 1996, pp. 433–471.

It was like walking into a candy store or a toy store and seeing all the toys on the shelves and being able to pick whichever we wanted to play with."[50]

Nirenberg was lucky that he worked at the National Institutes of Health where he could go on for years without applying for outside funding. Had he applied for support for the work, which led him to his seminal discovery, he is sure he would have not been granted support due to his lack of experience in the field. This is also why he was turned down when he applied to attend the Cold Spring Harbor Symposium in 1961.

By the time of the Moscow meeting, Watson had already been a Harvard professor for five years. He was successful in building up his laboratory, but his personal research achievements could not come close to the discovery of the double helix. The 1960s brought him the Nobel Prize and his great book successes. Four decades later, he could look back to his writings with pride.

What would be your longest ranging impact?

Probably my books. The discovery of DNA was just waiting to be made; it was not a difficult thing; any good chemist should've got the answer pretty fast. Rosalind Franklin was a physical chemist; she really wasn't a complete chemist. Pauling just goofed beyond any reason; it was crazy. Any good chemist should've found the structure of DNA. But *The Double Helix* was probably unlikely to have been written by anyone beside myself.

Then there is Molecular Biology of the Gene.

That also had a big impact.

There's a lot of chemistry in it. Did you learn that chemistry after the discovery?

Yes. You learn what you should know.

---

[50] Conversation with Marshall W. Nirenberg, unpublished records.

James D. Watson, fourth from left, at the celebration of the publication of the second edition of *Molecular Biology of the Cell* by Bruce Alberts, seventh from left; Dennis Bray; Julian Lewis, sixth; Martin Raff, third; Keith Roberts, eighth; and James D. Watson; and the first edition of *The Problem Book* by John Wilson, second, and Tim Hunt, first; also in the picture is Gavin Borden, publisher, fifth from left (courtesy of Tim Hunt, London; R. Timothy [Tim] Hunt won a share of the Nobel Prize in Physiology or Medicine in 2001 with Leland Hartwell and Paul M. Nurse "for their discoveries of key regulators of the cell cycle").

So it was not part of your training.

No, no, no. I liked natural history; I liked to be outside. I was interested in ecological successions. I thought I would be a naturalist and I didn't end up being a naturalist.

So your legacy is your books.

Yes, and probably this institution.

Watson realized early enough that he was not well versed in chemistry at the time of the double helix discovery and neither was Crick.

He noted that this ignorance "might have cost Francis and me the double helix."[51] Watson started remedying the situation about his chemistry as early as the fall of 1954. He signed up for Linus Pauling's course at Caltech, "Nature of the Chemical Bond."[52] By the time he completed his revolutionary major textbook *Molecular Biology of the Gene* in 1965, it showed no trace of his prior ignorance. At the turn of the millennium, *TIME* magazine asked Watson to write a tribute for Linus Pauling in which he graciously stated: "I most remember Pauling from 50 years ago, when he proclaimed that no vital forces, only chemical bonds, underlie life. Without that message, Crick and I might never have succeeded."[53] This is a graceful exaggeration in hindsight because Watson and Crick's quest for the structure of DNA had nothing to do with whether vital forces played a role in DNA replication or did not. On the contrary, Watson and Crick's discovery paved the way to the recognition of the pre-eminence of chemical

Erwin Chargaff at Columbia University in the late 1940s (courtesy of the late Erwin Chargaff) and in his home in New York City in 1994 (photo by the author).

---

[51] Watson, *Genes, Girls and Gamow*, p. 93.

[52] Ibid., p. 94.

[53] Watson. J. D., "Watson on Pauling." *TIME* 1999, 153, No. 12, March 29, p. 174.

bonding rather than the existence of vital forces in the life processes as was noted by Arthur Kornberg in his Nobel lecture in 1959.[54]

Talking about scientists' impact, scientific discoveries seldom have a long life for fame — except a few truly seminal ones and the discovery of the double helix is one of them — because if one scientist does not make this discovery another would soon. Chargaff referred to this when he argued against scientific autobiographies and what he said is instructive though he proved to be mistaken in the particular case about which he was writing. Chargaff wrote in his review[55] of *The Double Helix*: "There may also be profounder reasons for the general triteness of scientific autobiographies. *Timon of Athens* could not have been written, *Les Demoiselles d'Avignon* not have been painted, had Shakespeare and Picasso not existed. But of how many scientific achievements can this be claimed? One could almost say that, with very few exceptions, it is not the men that make science; it is science that makes the men. What A does today, B or C or D could surely do tomorrow." Chargaff was using this correct argument for the wrong example because the discovery of the double helix was one of those special events in science just as the book *The Double Helix* was a unique product of literature that only Watson could have written and no one else. Its freshness of exposition and complete lack of modesty were most unusual — until then — in such narratives. The book, which has been kept in print ever since has immortalized Watson in addition to the discovery.

The institution Watson refers to above is the Cold Spring Harbor Laboratory. Great scientists, such as Leo Szilard, Max Delbrück, Barbara McClintock, Alfred D. Hershey, and others have been associated with it. Watson was at first a researcher and a meeting participant here, then a trustee, and from 1968 director, later president, and after he stepped down from these functions, he is still the great old man at CSHL, although there is no want of titles; he is now the Chancellor of CSHL. Watson's wife, Elizabeth Watson has creatively contributed

---

[54] Kornberg, A., "The Biologic Synthesis of Deoxyribonucleic Acid." *Nobel Lectures in Physiology or Medicine 1942–1962.* World Scientific, Singapore, 1999, pp. 665–680, p. 668.

[55] Chargaff, E., "A Quick Climb Up Mount Olympus." *Science* 1968, 159, pp. 1448–1449.

to making the place not only into a citadel of science, but also an architectural gem, where over 200 years of architecture co-exist in harmony. Her pictorial history book about the place, *Houses of Science*[56] is a labor of love. It includes a series of essays by Watson, "Landmarks in Twentieth Century Genetics." Cold Spring Harbor Laboratory has developed graduate instruction, and there is now the Watson School of Biological Sciences. Watson has been interested in education, but would not call anybody his pupil.

Any pupils who carry on what you had started?
No. I had very good students but I don't ascribe their success to me.

You did not co-author papers with your students when you did not contribute to the work with your hands.
When I was in Indiana Luria didn't put his name on my paper and he didn't do the experiments. With time I decided that that was also what I was going to do. In a sense Luria's name shouldn't have been on the paper and I felt that my name shouldn't be on the papers that came out of my lab.

Because you didn't contribute to them manually?
Yes.

Doesn't this underestimate the intellectual contribution?
Maybe so, but people work best when they're working for themselves.

For Watson, Indiana University had a decisive impact on his career. In retrospect, Watson was glad that the California Institute of Technology (Caltech) had turned him down when he had applied to its graduate school. Although he realized that Bloomington, Indiana, was in the "intellectual backwater,"[57] H. J. Muller there had just

---

[56] Watson, E. L., *House of Science: A Pictorial History of Cold Spring Harbor Laboratory.* Cold Spring Harbor Laboratory Press, Cold Spring Harbor, New York, 1991.
[57] Watson, J. D., "Foreword" for Hargittai, I., *The Road to Stockholm: Nobel Prizes, Science, and Scientists.* Oxford University Press, Oxford, UK, 2002, p. vii.

received the Nobel Prize for the discovery of X-ray irradiation-induced mutations. Although Muller was not to be Watson's immediate mentor, there were others, most notably Salvador Luria, who most positively impacted his future career. This is well reflected in the Foreword he wrote for my book about the Nobel Prize, *The Road to Stockholm*,[58] which is reproduced in Appendix 4.

The brief part of our conversation above touched upon the important feature of the relationship between supervisor and young associates and about their respective fates. We start our comments from the latter, that is, how may Watson's approach to co-authoring papers with beginning researchers impact the career of the laboratory head. Benno Müller-Hill was for many years Professor of Genetics at the University of Cologne after he had done postdoctoral research in Watson's laboratory at Harvard in the mid-1960s. When asked about Watson's approach to not co-authoring papers with young associates, in 1999 he said:[59] "This is very good for the graduate student, for developing independence and responsibility, but may not be very good for the professor's formal productivity." Actually, Müller-Hill followed Watson's example for a decade in Cologne, but eventually had to give it up lest his applications for research support be turned down for lack of productivity.

What Müller-Hill experienced with Watson in the mid-1960s, though, was not what he might have in the mid-1950s. There must have been a turning point in between. In 1954, Watson's mentor and protector, Max Delbrück wrote a letter[60] to George W. Beadle in which Delbrück gives a devastating account of his pupil: "... He expected everybody to be concerned with his problems and could never bring himself to show the slightest interest in anybody else's problems. ... [Watson is] being ruthlessly egocentric in scientific matters. That goes to such ugly details as publishing prematurely, putting his name as senior author, accepting too many public lecture invitations, etc. ..." In contrast, in about a decade, Müller-Hill experienced

---

[58] Ibid., pp. vii–viii.

[59] Hargittai, *Candid Science II*, p. 128. (Benno Müller-Hill)

[60] Max Delbrück's letter of June 2, 1954, to George W. Beadle; Caltech Archives, as quoted in McElheny, V.K., *Watson and DNA: Making a Scientific Revolution*. Wiley, Chichester, 2003, Chapter 5 (p. 86), Ref. 38.

a diametrically different attitude by Watson as regards co-authorship with his associates and students. He would not be a co-author unless he manually participated in the work even though such an approach underestimated the intellectual contribution he might have also had.

Next, we turn to the problem of the beginning researcher about which the Danish Nobel laureate, Jens Christian Skou[61] had the following to say:[62]

"I am very sceptical about funding systems where you have to find all the money for your research from funds. To write applications for a grant which often is every second or third year is very time consuming. It is valuable in between to consider what you plan to do, but as it is impossible to foresee how your research will develop, you know that half a year or a year later you will probably not be doing what you wrote in your application. It is not only time consuming to write the applications, but also to evaluate them, meaning that highly qualified scientists spend a lot of time on evaluation, time which could be used better on research.

Second, you never know if you next time will receive support. That is why it is important to be able to present results at the next application, which may tempt to select problems you know can give results. It can be a hindrance for new thinking and for testing new ideas which may or may not give useful results, but which is so important for the development of basic research. Renewal of the grant also increases the pressure for publication, which may be too early.

Third, the funding system favors successful research. This, of course, deserves support. But with 50% to 55% of the applications worth supporting but with money only for 15% to 20% as it is in our country, this leaves little room for new

---

[61] Skou received half of the prize "for the first discovery of an ion-transporting enzyme, Na⁺,K⁺ATPase." The other half was shared by Paul D. Boyer and John E. Walker "for their elucidation of the enzymatic mechanism underlying the synthesis of adenosine triphosphate (ATP)."

[62] Hargittai, B., Hargittai, I., *Candid Science V: Conversations with Famous Scientists*. Imperial College Press, London, 2005, p. 451. (Jens Chr. Skou)

thinking. It is often the young who get the new ideas that move borders. It is very difficult for them with few or no previous publications to obtain money to work on their own ideas. Instead they must join teams who have money and work on the ideas of the head of the team. It is a hindrance for free research. Good research requires quality but also originality and engagement. And it stimulates the engagement to work on your own problem and be the author of the paper rather than to work on the problems of the head of the group and be one of many authors on the publication. With the present funding system it is important that heads of groups let the ideas of the single members develop and let them work independent on them, and not just focus on their own ideas.

Fourth, there is a tendency that the politicians use the funding system to direct the research. The money is put in subject earmarked boxes. In basic research it must not be the earmarked subjects, which determines the research, but the ideas of the scientists. To let the money determine the research subjects leads to lost possibilities or in the worst case to mediocrity."

Writing grant applications has become an industry. Today a young assistant professor at an American university is in a difficult position because he or she must excel in teaching while, simultaneously, writing grant proposals and carrying out independent research (for her this is even more difficult because this is also the time period when she may want to become a mother).

The old times can never come back yet it is of interest to see the problem of research support in some historical perspective. Erwin Chargaff liked to recall that the whole of funding for research goes back to Alexander von Humboldt, the German explorer, who founded Berlin University (today Humboldt University). Von Humboldt stressed that good teaching at university level can only be done if the person does research as well. The idea of Research University originated at Napoleonic times and not only in Berlin but even before in Giessen where Justus von Liebig, the great German chemist was its driving force. The institution of doctorate, doctoral studies, doctoral dissertation, eventually what we know of as PhD originated from the recognition that university instructors must learn doing research. In addition to training young people to do research, the other purpose of having PhD students is to assist the professor in doing the experiments for him.

In Chargaff's time in Germany, in the late 1920s and very early 1930s, the mechanism of getting funding was very different from today's practice. He was told that if he wanted to supplement what the university paid him and buy supplies for his research, he should apply to the funding agency, the Deutsche Notgemeinschaft for a grant. He made the application and got an invitation to see the chief of the agency, who was not even a scientist, but an orientalist. Chargaff had to tell him about the books he was reading and about his research plans, and had to explain some terms of his chemistry at the elementary level. He was told that he would soon hear from the agency and he did in three days, and had the grant. Chargaff gave his evaluation about this process:[63] "This is not as stupid as it may seem. If you get the right people they don't have to spend too much time on it. I find it silly when they call this proposal valuable and the other proposal not valuable. Most proposals are half so and half so because one doesn't know yet and if it is good it may still not work. So the peer reviews are completely useless except that they grow into old boys networks. I think the most important is to get the general behavior, the general way of thinking of the person rather than to decide that this is a marvellous problem. They are not marvellous except in lucky hands, in very gifted hands. The hands you can't look at in a proposal."

Linus Pauling was a great mentor although there were complaints that many of his projects for graduate students were routine tasks needed to be solved for Pauling's grand research schemes. For this, Pauling did not serve as role model for Watson, but in other aspects he did.

Somewhere you said that Linus Pauling was isolated by his greatness. Are you isolated?

I'm not great.

---

[63] Hargittai, *Candid Science*, p. 24. (Erwin Chargaff)

But are you isolated?

No. Linus didn't talk to anyone. He talked at people but he didn't talk with them.

He seemed to me accessible.

No one called him by his first name at Caltech. In America you use first names.

Do people stop you for a chat?

The scientists? Sure. It's totally different. Linus really wanted to do everything himself. He didn't want to learn from anyone. It finally destroyed him. His later career was pretty disgraceful; the Vitamin C and all that. He didn't want to accept data that disagreed with his ideas.

In his answer about his accessibility, Watson may have been mistaken. People do not stop him for a chat. When he comes to the cafeteria at CSHL to have a cup of coffee and have a look at the newspapers in the adjacent reading room, he is left alone. Everybody feels what Benno Müller-Hill and his fellow postdoc felt that it must be something important if you want to tell Watson about it. Watson describes his similar feelings when he remembers that during the last years of Alfred Hershey's life, as he was often passing Hershey's house, "I never thought I had something important enough for an excuse to interrupt."[64]

Watson and Pauling had much in common and no wonder that Watson tried to imitate Pauling, who was the dominant scientist for quite a while in 20th century chemistry. Pauling's chemistry leadership extended over decades by his evolving discoveries about the nature of the chemical bond and the discovery of the alpha-helix structure of proteins was one of its culmination points. He and his family were royalty at the California Institute of Technology just as Jim and Liz Watson have been royalty at Cold Spring Harbor Laboratory.

---

[64] Watson, J. D., "Alfred Hershey: A *New York Times* Tribute." In *We Can Sleep Later: Alfred D. Hershey and the Origins of Molecular Biology*, ed. Stahl, F. W. Cold Spring Harbor Laboratory Press, Cold Spring Harbor, New York, 2000, pp. xi–xii, p. xii.

In Watson's words, Pauling did not take easily differences in opinion, no matter how light-hearted,"[65] but these words would be at least as applicable to Watson at CSHL as to Pauling at Caltech.

With the Human Genome Project, more and more people may become informed about their fates and dangers of their genetic disposition. The new possibilities carry controversy with them, but that is something what Watson seems to thrive on. He is probably the best person to navigate in this intellectual mine field and he is doing it with gusto.

Intelligence and genetics: is this a taboo question?

For some people, yes. It's difficult to define intelligence while we don't really know how the brain works. I've always thought I have no mathematical ability and other people would say you have no mathematical interest and therefore you were not motivated to learn it. It's very hard to distinguish them. I think my brain works fast on what interests me. We don't know how we store numbers in the brain; we don't know any of these things.

But you are saying, "The misuse of genetics by Hitler should not deny its use today." What bothers me in this is that any demagogue says a lot of things that have truth in its roots yet the total is totally false. Couldn't we approach this whole problem [of the use of genetics] without invoking Hitler and Nazism?

No, because if I go out in the public, they say, "You're playing Hitler." If other people didn't raise it, I wouldn't but they constantly want to talk about the "eugenic past." The German disdain of the fact that the Greens have been so effective preventing DNA-based industry developing, at least they were in Germany, had a lot to do with the fact that genetics had become identified with Hitler. But genetics had been identified with Stalin too. Hitler used genetics arguments; Stalin denied its existence. You had two extremes. You have to

---

[65] Watson, *Genes, Girls and Gamow*, p. 31.

> know what's right and what's wrong. Hitler wanted the
> German race to be perfect so he wanted to kill off the imper-
> fect Germans while they were young.

The connection between intelligence and genetics seems to be losing
its taboo status recently. There is, for example, a treatise linking
above-average intelligence and above-average incidence of certain
degenerative diseases of the nervous system among Ashkenazi Jews.[66]

Watson has a non-compromising attitude towards eugenics in
general and towards the misuse of genetics by the Nazis in particular.
In 1997, Watson gave a keynote talk before a German audience of a
congress in Berlin of Molecular Medicine. The title of his presentation was "Genes and Politics." His talk followed the speech of a
German government official who discussed at length what the speaker
supposed to be the ethical issues of the human genome without even
hinting at Germany's past in eugenics and without mentioning
Nazism. Watson, on the other hand, after having discussed America's
poor record in eugenics, turned to the German experience, and talked
about details that may have been new to the 1500 German scientists
attending the congress.

I am quoting here from Watson's description of the event:[67]
"I suspect few of them had been consciously exposed to the gristly
details of how the German eugenicists gave wholehearted support to
the Nazi doctrine of race purity and their program to exclude anything
Jewish from future German life. At the end of my hour I emphasized
how bad it had been for the postwar German government to bring
back Professors Fritz Lenz and Otmar von Verschuer into German
academic life. Both had continued working on behalf of the Nazis
even after they knew the genocidal final solution to the Jewish problem
had commenced. Here, I had to say that the 1939 Nobel Prize winner

[66] See, e.g., Cochran, G., Hardy, J., Harpending, H., "Natural History of Ashkenazi Intelligence."
J. Biosocial Sci., 2006, **38**, pp. 659–693.
[67] Watson, J. D., "Afterword: Five Days in Berlin." In *Murderous Science: Elimination by
Scientific Selection of Jews, Gypsies, and Others in Germany, 1933–1945*, ed. Müller-Hill, B. Cold
Spring Harbor Laboratory Press, New York, 1998, p. 195.

Adolf Butenandt should be most remembered for his participation in the 1949 whitewash of Professor von Verschuer, then well known for wartime research using materials sent to him from Auschwitz by his former student Josef Mengele. Mildly reprimanded for 'a few isolated events of the past,' von Verschuer afterwards became the Professor of Genetics at Münster while Butenandt went on to lead the Max-Planck-Gesellschaft."

No wonder that Butenandt as President of the Max-Planck Society, prevented an investigation into the Society's past, the former Kaiser-Wilhelm-Gesellschaft, and such a probe took place only more than 50 years after the end of World War II. Even then it was done with great care not to upset people too much. When in 2001 I talked with Robert Huber,[68] Nobel laureate in chemistry (1988), he mentioned a very old scientist who still came to their Max Planck Institute from time to time and about whom it was known that he did experiments on children in the Nazi era. They did not know how to handle the situation and avoided even asking him not to come. Huber would not reveal his name and appeared very uncomfortable to come up with any detail of the story. He revealed to me that he had not yet worked out his own approach to how to view his relationship to Germany's past. Butenandt was Huber's boss at one time. Today, some of the chemistry institutes of Munich University are on Butenandt Street.

But let us continue the Watson conversation.

You are also saying that parents should have the right and the possibility to terminate a pregnancy.
Yes.

That means, you would stop this possibility before birth.
Yes.

---

[68] Hargittai, *Candid Science III*, pp. 354–367. (Robert Huber; some "sensitive" issues regarding German scientists' participation in experiments on humans were not included in the published conversation, at Huber's request.)

The plaque on the wall of the Eagle Pub in Cambridge, UK, commemorates Watson and Crick's announcement of having solved the secret of life on February 28, 1953, and was unveiled on April 25, 2003 (photo by M. Hargittai).

Not after birth.

Francis Crick gave a lecture in 1968[69] where he said you should only be declared alive, two days after birth. People have accused me of that remark, but it was Francis. Francis also said the state should not spend any money for medical care about people above 80. Now that he is 84 he would probably disagree. He said this when he was just about 50.

Would you subscribe to this two-days-after-birth definition?

Yes, I think it would be more compassionate and would generate more happiness in human society.

How do you set the limit and who sets it in the final account?

The fact that you have to set limits, that's true in whether you send someone to prison; there's always arbitrariness of rules.

---

[69] It was the Rickman Godlee lecture at University College, London, in October 1968, entitled "The Social Impact of Biology." A report about it appeared under the title "Logic in Biology," *Nature* 1968, 220, pp. 429–430.

If a child is born with just an almost hopeless form of cerebral palsy, I don't really think it's fair to anyone to say this child has to live through life of this absolute misery. I think what Crick said is something I believe would be humane. It would've been the way the Romans would've treated it; it's only with modern medicine. If something is going to be so grossly suffering, I wouldn't have wanted to be a burden on itself and on the others.

You don't strike me as a very passionate person but reading your book A Passion for DNA, *you come through to me as very passionate about this question.*

We should use this knowledge. People want to look forward to being happy and if you create a situation where it's pretty hard to be, I don't think it is right. There was this case of a fairly prosperous couple that had a child with cerebral palsy. They tried really hard and they ended up one day taking their child to a hospital or a police station, saying, "We can't handle it. We're incapable. It's beyond us." These weren't poor people, these weren't nasty people, but they were at their wit's ends.

The Nazis carried out human experiments. If there are data from those experiments, should they be used?

I'm not sure the data were usable. I think it's a philosophical question, not a practical question. I'm not sure there was much data of any value. They may have exposed people to high temperatures; they were monsters. But if the data still exist and they would save someone's life, I would save someone's life. But I'm not sure it's a very real question; I don't think there's much out there that would help us.

Making a stand about the Nazis' crimes may help us face other crimes.

I'm pragmatic; to save a child's life, I'd save it. But I don't think much is out there and I'm not going to worry about a philosophical question.

Here's a practical question. The Nazis killed a lot of mental patients whose brains are still being preserved in Germany.

I would've torn down the very building in Berlin, the Kaiser Wilhelm Institute; I would've just got rid of the f...ing place. But it's still there and they've put a plaque on it after a big fight. The Free University Faculty first didn't want a plaque on the building to say what it was.

So you would've destroyed the building.

I would've at the time.

But they are still there somewhere.

The buildings?

The brains.

I don't think they have any value.

Then why don't they destroy them?

I would. I'm not sure they are of any value any more than having Lenin's brain.

In 2003, I talked with Oleh Hornykiewitz,[70] Professor Emeritus at the Institute of Brain Research of the University if Vienna, Austria. He is most famous for showing that the lack of dopamine causes symptoms of Parkinson's disease in humans and for suggesting treatment with L-DOPA.[71] I brought up the topic of the many thousands of mental patients murdered in Nazi Germany whose brains were sent to various institutes ostensibly for research. Benno Müller-Hill, who has investigated the science in Nazi Germany, describes that a Professor Hallervorden received hundreds of brains. At least on one occasion, this professor went to one of the extermination centers, was present when the children, whose brains he wanted to examine, were being killed by gas, and showed the personnel how to take out the

---

[70] Hargittai, *Candid Science V*, pp. 618–647 (Oleh Hornykiewicz).

[71] The 2000 Nobel Prize in Physiology or Medicine was awarded to three scientists "for their discoveries concerning signal transduction in the nervous system." Oleh Hornykiewitz was not among the three awardees and 250 neuroscientists wrote an open letter protesting his omission.

brains from the victims' bodies fast, after the killing. After the war, Hallervorden lived the life of a respected German professor, a member of the Max Planck Society. The brains of victims were preserved for decades in Germany and also in Vienna. So I asked Hornykiewitz what he thought about this, and this is what he answered:[72]

"There had been discussions going on and on for many years about what to do with those brains. Most of the brains were kept stored in the Psychiatric City Hospital in the periphery of Vienna, where the victims had been housed and killed. But a few brains were also found in the Neurological Institute that does not exist anymore. Because of public pressure, two or three years ago the question was investigated again, and it was decided to put the brains to rest in a semi-religious ceremony in an honorary grave site at Vienna's Central Cemetery. But until the end, there were people who advocated keeping the material in a more visible form as a public memento, a reminder of the atrocities and the criminal and inhuman eugenics-madness of the Hitler regime.

Personally, I would find it unacceptable to use brains for research acquired in the way you described. When I started collecting brains, some 40 years ago, some pathologists asked me about the desired time interval after death. They were trying to be helpful to my research, offering me their special services. They thought that the interval could be shortened if I would so desire. My reply was and has remained, 'Follow your regulations and normal routine. Don't make any exceptions for me. I will not accept too long intervals, but until 24 hours it is fine for me.' During the four decades of my occupation with fresh postmortem brains, I had been variously offered extra-fresh brains, removed from the skull 30 or 50 minutes after death, whatever the definition of that was; or the possibility to receive larger amounts of fresh striatal tissue, up to 1 gram (!), taken during neurosurgical interventions on

[72] Hargittai, *Candid Science V*, pp. 646–647 (Oleh Hornykiewicz).

other brain areas. I turned them all down. Maybe I was wrong, but I felt such offers were going against medical ethics. You will even in vain look for a research paper of mine using brains of artificially aborted human fetuses, dozens of times offered to me. I believe in the very practical value of ethical barriers: they help to keep us human."

There have been conflicting opinions about the possible use of the findings of the so-called Nazi science. There is even a view according to which extracting useful information would somewhat alleviate the sacrifice of the victims. However, using such data is no way to honor their memory. Also, "scientists" who carried out experiments on humans, by doing so lacked moral compass and in that case it cannot be supposed that they had proper scientific compass either.

Do you think a genetic warfare is possible?
The Hutus killed the Tutsis.

I mean genetic warfare by means of DNA technology.
I don't know of any case of that. I don't see pills, which would selectively target one group and not another. We're pretty similar. All humans are separated by a hundred or a hundred and fifty thousand years, not very much.

François Jacob is not so sure that such weapons could not be created and neither is Matthew Meselson a long time expert on chemical and biological warfare.[73] To the question, "Do you think a genetic warfare is possible? Is it possible to develop biological weapons to kill a population selectively?" Jacob said in 2000:[74] "I don't know; for the present time I don't think so, but you have to wait to know more about the human genome. I expect surprises when the human genome becomes known completely. The notion of races may get a different interpretation, for example. We have to wait and see."

[73] Hargittai, I., Hargittai, M., *Candid Science VI: More Conversations with Famous Scientists.* Imperial College Press, London, 2006, pp. 40–61 (Matthew Meselson).
[74] Hargittai, *Candid Science* II, p. 94. (François Jacob).

Our next topic with Watson was not void of controversy although on a smaller scale.

About the Nobel Prize as an institution: When people call you the Einstein of biology, you are above that level.

It's very creepy.

What I mean is you are not concerned with the Nobel Prize any more and not only because you have it but also because your name is there regardless of any prize. The discovery of the double helix is part of science history. Mendeleev didn't get the Nobel Prize, although he could've, but it was voted down. However, his fame does not suffer from it.

Yes, sure.

But for the next layer of people the Nobel Prize is the ultimate recognition.

Yes.

How do you feel about it?

There're so many more people doing science so it would be nice if other Nobels came along.

There seems to be no competition, and when the Nobel Committees make some blunder, they appear to be re-writing science history.

Yes.

I'm not talking about Oswald Avery because it was such an obvious omission. But there are less clear-cut questions. One is about Isabella Karle who did important work in the applications of the direct method in X-ray crystallography. She was not included in her husband's 1985 Nobel Prize in chemistry, which he got with Herbert Hauptman; Hauptman and Jerome Karle co-discovered the direct methods.

She should've shared his prize. And Marianne Grunberg-Manago should've shared Ochoa's Prize. I think the Nobel

Left: Isabella Karle (courtesy of Isabella Karle, Washington, DC); right: Marianne Grunberg-Manago (courtesy of Florence Greffe and the Archives of the French Academy of Sciences, Paris).

Prize becomes much harder to fairly give with the increasing number of people who do science. The limit on three creates a lot of inherent injustice.

It was obvious that the discovery of the double helix was to be rewarded by a Nobel Prize. Why, do you think, was then a waiting of almost a decade for the Prize?

It wasn't obvious, but the Meselson-Stahl experiment showed that the strands did separate. I think also most biochemists and biologists thought in terms of proteins. They were really enzymologists. The fact that the information was as important, that the book was as important as the actors, that was not so obvious to many right away. Just the acceptance that genetics was in a sense more important than the downstream, that the information was the essence of life, wasn't accepted by a lot of people.

Even after the double helix was discovered.

Sure.

Left: Oswald Avery; right: Maclyn McCarty, James D. Watson, and Francis Crick in La Jolla, California, 1977, at the time Maclyn McCarty was given the First Waterford Biomedical Award (photo by Robert Smull — The Lensman Photography, San Diego, California; both photos courtesy of the late Maclyn McCarty).

Oswald Avery and his two associates, Colin MacLeod and Maclyn McCarty discovered experimentally that nucleic acids were the carrier of heredity. They published their findings in 1944,[75] but the impact of their findings was limited. Avery had been nominated repeatedly for the Nobel Prize for his studies in immunology, but never received the award. It was a rare admission of the Nobel Foundation in a book published years after his death that it was a mistake not to honor his epochal discovery.[76] Arguments that the full importance of the discovery could not yet be recognized are doubtful, however, because the Royal Society (London) as early as 1945 awarded Avery its highest recognition, the Copley Medal. Avery was not present at the award ceremony in London, where the President of the Royal Society, Henry Dale, himself a Nobel laureate said:[77] "Here surely is a change

---

[75] Avery, O. T., MacLeod, C., McCarty, M., "Studies on the Chemical Nature of the Substance Inducing Transformation of Pneumococcal Types: Induction of Transformation by a Desoxyribonucleic Acid Fraction Isolated from Pneumococcus Type III." *J. Exp. Med.* 1944, 79, pp. 137–158.

[76] *Nobel: The Man and His Prizes*, Third Edition. Edited by the Nobel Foundation and Odelberg, W. American Elsevier, New York, 1972, p. 201.

[77] Dale, H., Anniversary Address to the Royal Society, 1945, pp. 1–17, p. 2.

to which, if we are dealing with higher organisms, we should accord the status of a genetic variation; and the substance inducing it — the gene in solution, one is tempted to call it — appears to be a nucleic acid of the desoxyribose type. Whatever it be, it is something which should be capable of complete description in terms of structural chemistry." Considering omissions from the roster of Nobel laureates, François Jacob said in 2000:[78] "Avery's was an enormous discovery and I don't think there has been an omission comparable in magnitude ever since."

As for the Meselson-Stahl experiment, it has been labeled "the most beautiful experiment in biology," this label being quoted in the subtitle of a monograph fully devoted to its story.[79] The experiment used heavy isotopes to distinguish parental and progeny strands of DNA during its replication. It showed unambiguously that the two strands separated in the process. Nobody would have been surprised if the experiment had been judged worthy of a Nobel Prize, but this did not happen.

The limit of three awardees has played an important role in many cases, but not in Isabella Karle's or Grunberg-Manago's. Grunberg-Manago was a postdoctoral fellow with Ochoa from Paris, and her work substantially contributed to Ochoa's recognition.[80] In 1985, Herbert Hauptman and Jerome Karle shared the chemistry prize for the direct methods in X-ray crystallography. In both cases a third person could have been included. The direct methods in crystallography were Hauptman and Jerome Karle's joint invention. However, they did not go very far beyond having shown their validity; what was missing, was the practical applications. Several scientists contributed to making the methods widely available and in this Isabella Karle's

---

[78] Hargittai, *Candid Science II*, p. 90. (François Jacob)
[79] Holmes, F. L., *Meselson, Stahl, and the Replication of DNA: A History of "The Most Beautiful Experiment in Biology."* Yale University Press, New Haven and London, 2001.
[80] A footnote to this story is that two postdoctoral fellows arrived at the same time in Ochoa's laboratory and the other in addition to Grunberg-Manago was Irwin Rose. Ochoa offered them two topics and Rose chose what in retrospect proved to be of the lesser importance. Rose then went on to a solid but not especially remarkable career until in 2004 when — already in retirement — he was co-recipient of the chemistry Nobel Prize for the ubiquitin story.

Matthew Meselson and Franklin Stahl, who years before performed "the most beautiful experiment in biology," demonstrating the semi-conservative replication of DNA (courtesy of Franklin Stahl, Eugene, Oregon).

contributions were conspicuously important. She demonstrated the usefulness of the techniques in actual structure determinations. Jerome Karle was a member of the U.S. National Academy of Sciences and so was Isabella Karle, Herbert Hauptman was not before the Nobel Prize. The Karles were much in the limelight whereas Hauptman was a rather withdrawn person. Hauptman[81] would have not been surprised had the Karles received the Nobel Prize just by themselves or in combination with someone else. However, the general practice

---

[81] Hargittai, *Candid Science III*, pp. 314–315. (Herbert Hauptman)

of the Nobel committees is to go back to the initial publications and rely on them to make their decisions. According to such criteria, Hauptman and Karle must have been the unambiguous choice. Isabella Karle's inclusion would have been justified on the basis of later work, but the other names must have been considered as well.

Francis Crick mentions in his book *What Mad Pursuit* that when Arthur Kornberg began work on DNA replication, he did not believe in the replication mechanism Crick and Watson had proposed. It was only Kornberg's own experiments that finally convinced him that the two chains run in opposite directions. Kornberg himself told Crick this story.[82] We do not know whether Crick ever doubted the double helix structure, but Watson may have had doubts because he is known to have exclaimed that he had his first good night's sleep after he had learned about the single-crystal determination of the double helix structure of DNA, many years after their original publication.

The full stories about the Nobel Prizes can never be known because even the Nobel Archives, which become available for research after 50 years, contain nothing about deliberations and discussions. For the time element in the double helix award, it may have been an additional consideration that Crick and Watson were in much more junior positions to Max Perutz who was the head of the group in which the discovery was made. For W. Lawrence Bragg, the director of Cavendish at the time of the discovery of the double helix, and a great authority for the Swedes, Perutz was a clear favorite over Crick and Watson, and Perutz himself was in the race for a Nobel Prize with his hemoglobin studies. When Perutz and Kendrew finally obtained substantial results in 1959–1962 — far from the ones that they achieved eventually — they virtually immediately received the Nobel Prize.

In his biography of W. Lawrence Bragg, Hunter[83] discusses in some detail the 1962 Nobel Prizes in connection with Bragg's influence and input. He must have had other sources about Bragg's nominations

---

[82] Crick, F., *What Mad Pursuit: A Personal View of Scientific Discovery*. Basic Books, New York, 1988, p. 77.
[83] Hunter, G. K., *Light Is a Messenger: The Life and Science of William Lawrence Bragg*. Oxford University Press, 2004, pp. 226–233.

Max Perutz (left) and John Kendrew as drawn by W. L. Bragg in 1964 (courtesy of MRC LMB Archives, Cambridge, England).

than the archives of the Nobel Committees because they could not yet be investigated for the 1962 prizes. According to Hunter, Bragg wanted to be sure that King's College was represented in the Nobel Prize for the DNA structure in addition to Crick and Watson. Bragg did not make nominations for the DNA work until 1960 and by then Franklin had died. But even before, strong evidence suggests that Bragg knew only about Wilkins's contribution and not much about Franklin's to the discovery. Bragg made his nomination for the DNA work only after Perutz and Kendrew had achieved such results in the summer of 1959 that they could also be nominated for the Nobel Prize.[84]

In 1962, the Cambridge MRC unit triumphed with the chemistry prize for Perutz and Kendrew and the medicine prize for Crick and Watson, along with Wilkins from King's College, London. Perutz himself felt uncomfortable that his prize preceded Dorothy Hodgkin's who not only had been longer in the field (which in itself would not have been of concern), but whose achievements Perutz might have — rightly — considered superior to his own. Hodgkin received her Nobel Prize in 1964 "for her determinations by X-ray techniques of the structures of important biochemical substances."

---

[84] Ibid., p. 229.

Maurice Wilkins; Max Perutz; Francis Crick; John Steinbeck; James Watson; and John Kendrew at the Nobel Prize award ceremony in Stockholm, 1962 (courtesy of the late Lars Ernster).

One of the missing names from the Nobel roster is Erwin Chargaff's. He made seminal discoveries, but somehow failed to bring his observations of the base ratios to a conclusion. He was not very good in popularizing his achievements either. Chargaff received many awards and so did Oswald Avery. Both Avery and Chargaff were not quite prone to recognition whereas other people are more disposed to it.

Have you seen Erwin Chargaff lately?

I saw him about six years ago at some National Academy function in New York. He is obsessed by what many Central Europeans think that truth comes from words. I don't think it does.

But you also find language very important.

I think finding the right word always conveys reality so much better; having a rich vocabulary enriches your life but philosophy is just distraction from reality.

If you could meet any scientist from the past, who would that be?

There are obvious ones, Mendel, Darwin. More recently I was curious about Calvin Bridges, the great *Drosophila* man at Columbia. There was Morgan and his three students, Bridges, Sturtevant, and Muller. I met Muller, he was one of my teachers at Indiana and I met Sturtevant, and Bridges seemed to be an interesting person. The most influential biologist of all time is Darwin, just overwhelming.

Is there any consequence of Asilomar still in effect?

This is different in Europe and the United States.

Erwin Chargaff was one of the two scientists for whom Avery *et al.*'s discovery about the nucleic acids being the hereditary material was a seminal event (the other was Joshua Lederberg[85]). The already successful Chargaff changed his research direction immediately upon learning about Avery's discovery, and devoted himself and his laboratory at Columbia University to the study of nucleic acids. He made two principal discoveries. One was that DNA was far from a dull molecule. He disproved the long-held view that DNA consisted of a regular and repeating chain of the four nucleotides; rather, the four nucleotides and their relative proportions differed considerably in the DNAs of different organisms. In other words, DNA was organism-specific. The other finding concerned the relative proportions of the four bases in DNA. He recognized a pattern in that the amount of adenine always equaled the amount of thymine, and the amount of guanine always equaled the amount of cytosine. Perhaps sensing the enormity of the consequences, Chargaff appeared reluctant to believe his observation. His hesitation is displayed in his report:[86]

"The results serve to disprove the tetranucleotide hypothesis. It is, however, noteworthy — whether this is more than accidental, cannot yet be said — that in all desoxypentose nucleic

---

[85] Hargittai, *Candid Science II*, p. 37. (Joshua Lederberg)

[86] Chargaff, E., *Heraclitean Fire: Sketches from a Life before Nature*. The Rockefeller University Press, New York, 1978, p. 93 (quoting from Chargaff's *Experientia* paper in 1950).

acids examined thus far the molar ratios of total purines to total pyrimidines, and also of adenine to thymine and of guanine to cytosine, were not far from one."

To these words expressing skepticism in 1950, he added the following three decades later:[87]

"For a long time I felt a great reluctance to accept such regularities, since it had been impressed on me that our search for harmony, for an easily perceived and pleasing harmony, could only serve to distort or gloss over the intricacies of nature. Many people in the past had attempted to find unifying formulations for the proteins and other natural high polymers, just as the nucleic acids had been considered as tetranucleotides, because they were built of four nucleotide constituents."

When I talked with Chargaff in 1994, I asked him an inaccurate question, which only displayed my ignorance at the time. I asked him whether base-pairing was his most important contribution.[88] I should have known that he did not discover base-pairing, only base equivalence, and he should have corrected me. Instead, he thought for a long time and answered distinctly, "Yes, it was."

Of course, it could have been an oversimplification for the sake of an apparently ignorant interviewer, but his response to my next question made it clear that if my question was superficial, his answer was well thought-through. I asked him whether he felt the importance of this finding at the time when he made it and this is what he said:[89]

"Yes, I did. I've often been asked about the double helix and this is related to it. In a way the double helix is a gimmick because there is no evidence that it occurs in the body. It

[87] Chargaff, *Heraclitean Fire*, p. 94.
[88] Hargittai, *Candid Science*, p. 24 (Erwin Chargaff).
[89] Ibid., pp. 24–25.

occurred in X-ray diagrams. What was important in Crick and Watson's work was that they proposed something that was *double*. The shape of helix is not so much of consequence. The base-pairing seems to be a novel law of nature that had not been dreamed of because the amino acids and the fatty acids, none of them show this gregariousness, this steric fit. Therefore, I always say that of the three Chargaff rules, as they are called, the most important is the one stating that six amino equals six oxo. That is true not only for intact DNA but also for each of its separate strands, whereas the other rules are not true, and it is true also for the complete RNA in the body."

In contrast, his description of his findings in his *Heraclitean Fire* 1978 is more accurate:[90]

"The regularities of the composition of deoxyribonucleic acids — some friendly people later called them the 'Chargaff rules' — are as follows: (a) the sum of the purines (adenine and guanine) equals that of the pyrimidines (cytosine and thymine); (b) the molar ratio of adenine to thymine equals 1; (c) the molar ratio of guanine to cytosine equals 1. And, as a direct consequence of these relationships, (d) the number of 6-amino groups (adenine and cytosine) is the same as that of 6-keto groups (guanine and thymine)."

However, two pages before in the same book[91] he describes the circumstances of his discovery of base equivalence in the quiet of his office at Columbia University when he asked himself, "What would happen if I assume that DNA contains equal quantities of purines and pyrimidines?" As he was examining the numerical data, in Chargaff's words:[92] "there emerged — like Botticelli's Venus on the shell, though not quite flawless — the regularities that I then used to call

---

[90] Chargaff, *Heraclitean Fire*, p. 96.
[91] Ibid., p. 94.
[92] Ibid.

the complementarity relationship and that are now known as base-pairing." This was the point where Chargaff misled himself in hind-sight, but he fully realized — however sarcastically he expressed it — that he "missed the opportunity of being enshrined in the various halls of fame of the science museums."[93]

The weight of Chargaff's observation from the available evidence can only be appreciated by looking up his raw data in which there was substantial scatter, and yet he came out with the 1:1 ratios. It was a daring conclusion. He noticed a pattern and did not refrain from com-municating it where most people might have seen just mere numbers.

Chargaff's generally known and noted bitterness has been inter-preted by most as his dissatisfaction of recognition of his scientific results. It is true that he never received the Nobel Prize though was awarded by numerous other important prizes. In his review of *The Double Helix*, J. Desmond Bernal[94] discussed the road to the discov-ery and Chargaff's role in it. According to Bernal, the complementary self-replicating character of the double helix was almost self-evident after Chargaff had given the clue to it by having observed that the sum of the pyrimidine bases was equal to that of the purine bases. From this it followed that "they could occur in natural paired bases, the pairs being held together by hydrogen bonds."[95] However, by stating this, Bernal overextended the inferences Chargaff had drawn from the equivalence of the pyrimidine and purine bases. Thus Bernal exaggerated what need not be exaggerated — Chargaff's contribu-tion. However, if even Bernal could do so, then no wonder that so could Chargaff.

Chargaff's name came up early in our conversation with Francis Crick on February 7, 2004. He found it strange that Chargaff did not discover base-pairing in the light of his observations on the base ratios in DNA. Crick quickly added though that Chargaff's mind might have not wandered towards pairing because he, that is, Chargaff was thinking in terms of one chain rather than two. In a single chain,

---

[93] Ibid., p. 98.
[94] Bernal, J. D., "The Material Theory of Life." *Labour Monthly* 1968, pp. 323–326.
[95] Ibid., p. 325.

nothing would prompt one's thinking towards pairing. Once two chains need be considered, pairing enters one's thinking more naturally. Crick seemed careful not to use the word helix at this point, as if placing himself into Chargaff's position.

Once the idea of two chains or helices came up, base-pairing was more probable to be thought of. That it was not trivial is witnessed by the fact that Crick and Watson did not think of base-pairing either until very late in the story of the discovery when the complementary arrangement came up. Even then, as Watson was pairing bases, initially he was seeking correspondence between like bases.

Crick gently reminded me that solving the problem was less straightforward than I might have thought. Complementarity could have also been accomplished between like bases. When they started thinking about pairing bases though, whether like to like or between different ones, the solution was quickly found. As Crick was talking about finding base-pairing, he distinctly spoke about "our" and not just Watson's findings. According to Crick, Watson did not even want to believe in base-pairs initially.

It is also of interest to see Chargaff through the eyes of a close observer, his son.[96] He moved away from his father's profession as far as he could. He trained as a hospital administrator, then became a policeman, later detective, who first worked on armed robberies of businesses and then on murders. He certainly sensed his father's pessimism and negative outlook on virtually everything. He knows about the importance of Chargaff's work, not only in relation to nucleic acids, but also about his earlier work on the mechanism of blood clotting, which saved a lot of lives during the Second World War. He thinks that his father is little known in the United States and is more famous in Europe. A sign of this was when they had a minute of silence on national radio in Austria when his death was announced. According to the son, his father's bitterness came from his feeling that Crick and Watson got what should have been his Nobel Prize. He openly shared what he knew with them, and they took his years and years of work and just put the final finishing touch on it. He compared this

---

[96] E-mail exchange with Thomas Chargaff in the spring of 2006; unpublished records.

to decorating a Christmas tree and then someone else putting a star at the very top and announcing that they had decorated the tree! He is also familiar with another version of the story according to which his father was slow, methodical, careful, and stopped short of drawing the final conclusion.

In light of the above discussion then it is unexpected that Watson himself speaks about "Erwin Chargaff's *base pairs*" (my italics) when decades later, in *Genes, Girls and Gamow*[97] he mentions his and Crick's lack of knowledge of chemistry at the time of the double helix discovery. Watson stresses that it was crucial that the Caltech-trained chemist, Jerry Donohue set them on the right course about the two possible forms of the nucleotide bases in DNA and thus led them to recognize "Chargaff's base pairs." Thus in this case Watson himself makes the mistaken claim for Chargaff. But there have been other claims, too.

In 2003, Maurice Wilkins published his autobiography with a telling title, *The Third Man of the Double Helix*.[98] In this book, he, too, makes the claim of having recognized base-pairing. The time was January 1953, according to Wilkins, when he showed the visiting Watson the famous photograph of the B pattern of DNA. Wilkins then adds: "I felt I must tell him what I had been thinking about base-pairing in DNA. I had only recently had an idea and, because I respected him as a scientist and knew he had thought a great deal about DNA structure, I was eager to discuss the idea with him. In the event, however, I got no further than saying 'I think Chargaff's ratios are the key to DNA structure,' and Jim said 'I do too' before he hurried off."[99] Wilkins discusses the history of his developing the idea of base-pairing and spells out that he had thought before — he would hear about it from Crick and Watson — that "The bases would be hydrogen-bonded together in line with Chargaff's 1:1 ratios."[100] Crick thought that Wilkins speaks about base-pairing in his book as if

---

[97] Watson, *Genes, Girls and Gamow*, p. 93.
[98] Wilkins, M., *The Third Man of the Double Helix: The Autobiography of Maurice Wilkins*. Oxford University Press, 2003.
[99] Ibid., p. 198.
[100] Ibid., p. 200.

he knew more about it than he could have and did at that time. Crick was sure that Wilkins did not have the idea at the time.[101]

The other topic that needs some commentary here is *Asilomar*. There were two Asilomar conferences, in 1973 and 1975, about the hazards of experiments with gene manipulation when the need for self-regulation by the scientific community arose. This is how Paul Berg, one of the principal players summarized the events and their consequences in 1999:[102]

"When we began to do these experiments [genetic manipulation], people worried about 'Would we be making bacteria more infectious?' In other words, if I take a gene from a cancer virus and I put it into a bacterium and that bacterium can

Paul Berg, whom some call "the father of genetic engineering" and who was co-organizer of the Asilomar meeting, at Stanford University, 1999 (photo by the author).

[101] Hargittai, *Candid Science VI.*, pp. 2–19. (Francis Crick)
[102] Hargittai, *Candid Science II*, pp. 166–168. (Paul Berg)

live in your intestine, does that create a big risk that you'll develop intestinal cancer? We didn't know the answer. If you take genes from organisms that are not sensitive to certain antibiotics and you put them into other bacteria that are sensitive to the antibiotics, do you change them to become resistant to the antibiotics? And the answer to all those questions is yes. So at that point we said, 'Wait a minute, perhaps we should stop doing these kinds of experiments and determine if there is a real hazard or risk in doing these kinds of experiments.' We certainly did not want to send bacteria out into the environment that could infect people, cause cancer or whatever. At the time, however, we had no way to determine whether the risk was real or imaginary.

As soon as it became clear from these experiments that you could move genes to and from all kinds of organisms, people said, 'Are we going to create some monsters?' I was asked by the National Academy of Sciences to convene a group that would give advice to the Academy of what to do with this new science. This small group published a letter in the journals *Science* and *Nature*, which became known as the 'moratorium letter.' The 'moratorium letter' suggested that everybody around the world should stop doing these kinds of experiments until we could meet and determine whether they posed a risk or not. This letter had a profound effect all over the world, and every country issued it to all its scientists working in this field, 'Stop doing these experiments.' In that letter, we recommended that there be an international conference where the scientists working in this field would come together and try to arrive at a judgment about whether these experiments were risky.

As a result, a now famous meeting called the Asilomar Conference was held in 1975. Asilomar is a big conference center on the Monterey Peninsula near Carmel, about a hundred miles south from here. The conference brought together 150 people from all over the world — from Russia, China, Europe, and the United States — all people who wanted to use this technology of genetic manipulation. The purpose was to ask the question, is it safe? And if it's not safe, what can we do to reduce the risks so that we can do the research. At the conference, we decided that we didn't have enough information to determine whether it was really safe or not. What we could say was that there are some experiments for which there is almost zero risk, and here is a group of experiments for which there are possible risks, and another series of experiments for which we think there's serious risk. We recommended working out a set of instructions and guidelines which everybody would accept: These kinds of experiments could be done in any laboratory, these kinds of experiments would be done only in special laboratories, and these kinds of experiments would be done in only the most secure laboratories. The guidelines essentially regulated research for about 12 years, and because they were not set up by legislation, they were only recommendations, they could be changed.

As we learned more and more and found out that experiments we were worried about were perfectly safe, we could move them from the high-risk category to the low-risk category. Today there's almost no restriction on the kinds of experiments and molecules that you can work on anywhere. The work was going slowly at the beginning because people had to build special laboratories for it, but as it became clear that it was safe, the work expanded and exploded. Today I am quite confident that there is no risk to any of the kinds of experiments that we have been doing over the last 20 years."

Watson found the Asilomar meetings important, but thought that they produced self-inflicted wounds for science. Rather than demonstrating the scientists' responsible behavior to the public, these meetings enhanced distrust in science. Of course, everything may look different in hindsight, now knowing that the consequences of gene manipulation have been kept under control in the laboratory. Almost two decades later he summarized his stand on Asilomar in the following way:[103] "We did suffer initially from unfounded but widespread fears that the recombinant DNA revolution might create altered life forms that could threaten human existence, either by their ability to cause disease or by drastically upsetting the world's ecological balance. At first, unnecessarily harsh restrictions were imposed on our ability to work with recombinant DNA. But we fought back and by 1979, we were not effectively restricted in the United States in our ability to work with human DNA. Many of the fear-mongering scientists had long been known for one reason or another to just *dislike DNA*" (italics added). It is hard not to notice here that his onslaught is being directed not against people who criticize certain experiments or procedures, but who "dislike DNA." This was a sign that Watson was increasingly identifying himself with not just the double helix structure of DNA, but with DNA itself (see this again in the Epilogue).

Watson co-edited a volume, which tells the story of Asilomar.[104] When Watson refers to the difference in the consequences of Asilomar between Europe and the United States, it is that there is much greater apprehension in Europe for genetically modified food than in the United States and my next question followed this up.

Genetically modified food? You may have seen the European resistance to it.

They're trying to bring it over here. I'll give you Prince Charles' latest [anti-GMF] statement. You can have fun with that. They've asked me to reply to it because the British people don't find it easy to attack him. He's totally misguided but he dislikes modern architecture, too. He's caught in the past and would like to go back to when kings were important. He's harmful to the people. If Charles were king, I would be a republican.

---

[103] Watson, J. D., "Looking Forward." *Gene* 1993, 135, pp. 309–315, p. 312.
[104] Watson, J. D., Tooze, J., *The DNA Story: A Documentary History of Gene Cloning*. W. H. Freeman and Co., San Francisco, 1981.

Aaron Klug with the tobacco mosaic virus model at MRC LMB in 2000 (photo by the author).

A balanced discussion of GMF was given by Aaron Klug in his Anniversary Address as President of the Royal Society in 1998:[105]

> "The latest example to disturb the public is the use of genetically modified plants for food use. Here there is a legitimate debate on the benefits versus the risks of this technology. The latter are slight for humans, but with possible dangers to biodiversity and the environment. These last questions can be

---

[105] Klug, A., "Anniversary Address by the President." *Royal Society News*, London, December 1998, pp. 1–6, pp. 5–6.

addressed only by monitoring, and by experiments and field trials. The benefits are likely to come in the future. When one considers that the world population is expected to reach eight billion by the year 2020, it is clear that new approaches in crop and animal husbandry will be needed to feed such a burgeoning population. Strains genetically engineered to produce bumper yields or grow on marginal soils would reduce the need to destroy rainforests or drain wetlands for farming. There is a possible green agenda here, waiting to be written.

But the general effect of the introduction of GM plants has been unsettling. The public asks why we need GM plants at all, why add to our traditional foods. Are they an unnecessary and dangerous novelty foisted on us by multinational corporations? Moreover, in one well-noted case, bioengineered Soya products have been mixed with normal ones, allowing consumers no choice and increasing public anxiety.

Above all, GM foods are seen by some as an unwarranted interference with nature. This time, some would say, science has gone too far. I would rather say, too fast to absorb. The fears of the public are legitimate, and it is up to us scientists to convey a real understanding of the issues involved. We should remember that scientists and the public work from different cultural viewpoints. Research scientists are used to an uncertain world where knowledge is incomplete, and where trials and experiments are used to reduce the uncertainty. So I return to the general theme of the necessity for engaging in debate and explaining the importance of science to society, and, notwithstanding the problems it can bring in its train, why we continue to go forward."

It seems to me that the world is getting increasingly centralized. For science, the northeast and the southwest of the United States and Cambridge, England are at the top. Of course, I'm oversimplifying and exaggerating but when Manfred Eigen has something important to say he goes from Goettingen to La Jolla to talk about it and it's not the other way around.[106]

---

[106] Hargittai, *Candid Science III*, pp. 368–377. (Manfred Eigen; he became Nobel laureate in 1967)

We should keep science from being too centralized. It's partly government policy. French science has always suffered from Paris being too important. Luckily we have our Federal system by which the money is distributed partly on the basis of geography. That's good. You can do very good DNA science in 50 parts of the United States. America is prosperous and is putting even more money into science. Europe is not following in the same way. When I went to Cambridge, England, in 1951, it was the best university in the world.

You have been critical of the UK MRC Laboratory of Molecular Biology by saying that its conditions were so good that people were reluctant to leave and the impact of the Laboratory was diminishing.

People don't want to leave Cambridge; there should be greater incentives for leaving.

Comparison of expenditures on Research and Development in percentage of the Gross National Product among the three main units in the Western world is revealing. As of 2005, the United States (2.6%) and Japan (3.2%) have overtaken the European Union of the 25 member states (1.9%). The latter has a long way to reach its stated goal of 3% by 2010, with the Scandinavian and a few other countries already far exceeding it.

The great and continued success of the Laboratory of Molecular Biology in Cambridge has intrigued many over the years. On April 26, 2003, Sidney Brenner gave the closing talk at the one-day program of lectures at the MRC LMB devoted to the 50th anniversary of the discovery of the double helix. About LMB, he said that it was not only ideas and experiments; an important factor of its success was that it was also developing the necessary tools to carry out research, such as X-ray diffraction apparatus, apparatus for electron crystallography, the experimental setup for sequencing, etc. People found collaborative efforts very useful, but they had to form collaboration themselves; there was no institutionalized effort to this end. Brenner quoted Crick's opinion that you can't make people collaborate, but surely you can stop them.

Left: Six Nobel laureates — present or former associates of MRC LMB — James D. Watson; Max F. Perutz; César Milstein; Frederick Sanger; John Kendrew; and Aaron Klug (courtesy of MRC LMB Archives).

Brenner made reference to Watson's remarks in his brief contribution on the previous day, at the conclusion of the general symposium. Watson mentioned that the thing to do was to get into a new research area from the very beginning, and Brenner wondered how one could know when the beginning was and not get into it before. On the other hand, Brenner disagreed with the notion that we should have only big labs in the future because some of those big labs, like CERN, are just factories, and in a big lab, the members may be working on details without knowing what the whole problem is. Brenner stressed that one of LMB's secrets was that it early on recognized the power of computers in handling complex information.

Of course, Watson felt a tremendous nostalgia for Cambridge and for his partnership with Crick in Cambridge. Eventually, Crick also felt being forced out of Cambridge when retirement age approached and he moved to California. When Watson returned to Cambridge in 1986,[107] and toured the places where they had been together, Crick's absence must have been a painful reminder of his own transience. Institutions, like Cambridge University — and Cold Spring Harbor

---

[107] Watson, *Genes, Girls and Gamow*, p. 3.

Francis Crick and James D. Watson at the Backs in Cambridge, England, 1950 (courtesy of James D. Watson, Cold Spring Harbor, New York).

Laboratory just as well — were more lasting than individuals, however famous they may have been.

> *Quite a lot of money is being spent on popularizing the Human Genome Project, its ethical, social, and legal aspects. What more could be done than what's already being done?*

Right now it's not the misuse of genetics, rather the disuse of it, that we're not using it. We're not screening people for cystic fibrosis.

Why not?

There are two reasons. People are afraid to look into the future and they don't want to discover that maybe they're carrying something that could cause them harm in the future.

Left two photos: 19-20 Portugal Place in Cambridge, England, the Cricks' former home with the helix above the entrance. Right: Plaque commemorating the discovery of the double helix on the wall of the courtyard of the old Cavendish Laboratory in Cambridge, England (all three photos by the author).

That's partly it and partly, essentially, the Right to Life movement doesn't want genetic knowledge to be used; tests, which
might lead to abortion, are thought to be inherently bad. I
was speaking to a really bullshitting rabbi last night, an awful
character. He said there was something morally wrong about
abortion. I said it was unpleasant but if it'd prevent a woman
from future unhappiness, I think it's a good thing. There are
religious things. Then, there's the left-wing dislike of genetics
because they don't like to see the fall of society in bad genes
but in capitalism. I was just in Jerusalem and I saw the zealots.
They were even worse 2000 years ago than they are today.
They killed for God more intensely then than they would
today, so it's not new. The Third Temple could be destroyed
because of the zealots. The Third Temple is modern
Jerusalem and these people can create a situation in which
they destroy themselves if they refuse to grant any rights to
the Palestinians. Saying that Jerusalem is a Jewish city is crap.
There are so many religions there that it is tolerance that let
people survive, not proclaiming the truth. It's scary when you
see what these people are willing to die for. I'm worried that
the Jews can destroy themselves again by being zealot. I saw
a lot of Dick Feynman back when I was at Caltech and we
both got a letter from a rabbi asking for our religious views.
Since I wasn't Jewish I didn't write back but Dick wrote back,
saying, what a pile of crap.[108] As I was telling someone, the
Catholic Church would've been in a different place if instead
of imprisoning Galileo they had made him a saint. The true
miracles in the world come from science, new ideas, the cur-
ing of disease, smallpox. Basically the scientists are the people
who do the impossible. The only people who do the impossible are those who save people, and that comes from knowledge. It doesn't come from prayer. The important thing is
knowledge, not prayer.

---

[108] In *Genes, Girls and Gamow*, p. 104, referring to the same episode, Watson writes that he
wrote back saying that he had no interest in religion.

Every American coin says, "In God We Trust."

That's part of our heritage. The only thing I trust is that the human being is a social creature. We're born to cooperate with each other and if we don't, we're in deep trouble. Our values come from our being a social species.

Do you have a message?

Because of science, people have better lives today than before and in 2010 life will be yet better. But I don't think too far ahead. We are where we are because we use knowledge and we have culture, which we can store now. There will always be uncertainty. The thing that finally could do us in would be some form of disease that we're not prepared for.

The main thing is you don't kill someone because you dislike him. I think genetics will help people more but looking into the future is scary. I don't want to know when I go to the doctor what he'll say. But, on the other hand, you may look into your future and say that your blood pressure is high and we can lower it now and you don't have to die of high blood pressure the way you did 50 years ago.

Do you have high blood pressure?

Maybe slightly high but my father died of a stroke early.

To change the topic, a person totally irresponsible in some way but in others the most remarkable person of the century was not Einstein but Szilard. Leo Szilard always thought three or four steps ahead. If A meant B and B meant C, Leo would worry about E. Most people never could think ahead that far.

Then one thing puzzles me. Szilard had made this remarkable conversion from physics to biology. When Nirenberg made his discovery in 1961 he asked Szilard to sponsor his paper in the Proceedings of the National Academy of Sciences. *Szilard at that time had his headquarters in the lobby of the Dupont Hotel in Washington. Nirenberg spent most of a day telling Szilard about his discovery and in the end Szilard declined to sponsor it, saying that he was not a biologist.*

Szilard was recently elected member of the National Academy of Sciences of the USA and this gave him the right to sponsor papers by non-members in the *Proceedings* of the Academy. So Nirenberg had asked him to sponsor the first papers about breaking the genetic code. When Nirenberg called him, Szilard asked him down to the Dupont Hotel. Szilard was deeply involved at the time in questions of defense. This was during John F. Kennedy's presidency when Szilard found there was some openness for his wisdom on the part of the Kennedy Administration. People from the Pentagon and from official Washington came to see him and he would confer with them in the hotel lobby. When Nirenberg came, Szilard asked him to explain his work to him. Nirenberg spent the whole day with Szilard explaining to him what they had done and what the implications were. During their meeting, there was a stream of people that kept interrupting them. At the end, Szilard told Nirenberg, "It's too much out of my field. I'm sorry, I can't sponsor it."[109] This was an unexpected ending of this charming story because Szilard was famous for his foresight and, at the time of this meeting he had been much involved in biology. Independent of this outcome, Nirenberg and Szilard had a cordial relationship and Nirenberg saw Szilard on numerous occasions when Szilard visited the National Institutes of Health.

> He [Szilard] made a mistake, a bad mistake. A year before, I'd gone to Leo at the Memorial Hospital and told him about the evidence on messenger RNA and about the ribosome. It was late March or the first day of April and Leo said he didn't believe it. Leo was consumed by that time with the bladder cancer. To me he was very interesting; Einstein was pretty boring; Leo was much more complicated than Einstein was. Was it one high school in Budapest that produced Szilard and the others?
>
> *That's a myth. There were several high schools. There was also a strong Jewish upper middle class. Szilard, Wigner, von Hevesy,*

---

[109] Hargittai, *Candid Science II*, p. 140. (Marshall W. Nirenberg)

Teller, von Neumann, von Kármán, they didn't all go to the same high school.

Was Hevesy Jewish?

Yes.

Just part Jewish? Didn't they acquire the title by marriage, von Hevesy, von Neumann, von Kármán?

No, their parents had got the title. Austria-Hungary was a rather enlightened country around the turn of the century and Jews, even non-converted Jews, could be elevated to nobility, and they were.

Leo was extraordinary. Being in England and worrying about the chain reaction and trying to get people not to publish and getting the purified carbon [graphite], which the Germans didn't do. Leo realized that impurity could f... up the system, and Heisenberg didn't. The thing worked. Fermi would've never got us the thing; it was Leo.

By this point in our conversation, its character had changed; something must have clicked along the way. At times we assumed reversed roles in who was asking the questions and who was answering them. Watson showed genuine interest in Hungary and its history, and, in particular, in its intellectual life.

Fin de siècle Hungary and especially its capital, Budapest, were fast developing. In addition to excellent denominational schools a network of state schools appeared, the most famous being the Minta (Model) high school.[110] Theodore von Kármán, Michael Polanyi, Edward Teller, Peter Lax, and other luminaries were Minta pupils. Leo Szilard graduated from another, lesser known state high school. Nobel laureate physicist Eugene P. Wigner, John von Neumann, and John Harsanyi (Nobel Prize winner in economics) graduated from the Lutheran high school and the Nobel laureate chemists George

---

[110] More about this, in Hargittai, I., *The Martians of Science: Five Physicists Who Changed the Twentieth Century.* Oxford University Press, New York, 2006, pp. 11–17.

Liz (on the left) and Jim Watson with Magdi Hargittai at the Minta Gimnázium ("Model High School"), Budapest, 2000 (photo by the author). The plaque commemorates the founder of the school, Maurice von Kármán, the father of Theodore von Kármán.

von Hevesy and George A. Olah from the high school of the Piarist Order.

Fermi's and Szilard's names are, of course, forever linked in Einstein historic letter sent to President Roosevelt about the necessity of the United States paying attention to the possible development of nuclear warfare. They also worked together in creating the world's first nuclear reactor, called atomic pile at that time. Apart from the fact that Fermi was one of the greatest physicists of the 20th century, the two were very different. For Fermi, science was his life. Szilard lived in two worlds in which science was in one, and it was of lesser importance. He set out to save the world and wanted to use science to advance this goal; this was his other world. The two scientists being so different it is remarkable that they could and did cooperate in crucial historic moments.[111]

---

[111] Ibid., pp. 188–195.

There are some that question the importance of his contribution to the project.

He [Leo Szilard] was socially inept; he didn't make people feel good.

If a person like Leo Szilard would show up tomorrow at your doorstep, would you be able to recognize his potential, would you give him support, a place?

Possibly not, because he was always two steps ahead of everyone else. When you go beyond, it threatens people. Crick and I, we threatened the biochemists tremendously. This was part of why we didn't get the Nobel Prize. We were threatening. We were slightly ahead of time.

There may be the equivalent of Leo today, somewhere around, worrying about how the brain works, which is to me the big and interesting problem. I was very lucky. Occasionally I say I have a Szilard idea. It's good.

Leo could also be bizarre. He wouldn't flush the toilet and this is how George Klein discovered his bladder cancer. George Klein and Benno Müller-Hill are rare examples in science who do everything for science and jealousy doesn't motivate them. They're the people I admire.

There is also a view according to which the Einstein letter to President Roosevelt, engineered by Leo Szilard, hindered rather than accelerated the development of the American atomic bomb because it deemed the not yet existing program to wander in the bureaucratic maze. It could have been started much sooner, had it been initiated by people better versed in American conditions. It finally happened due to the impact of British scientists. If properly initiated, the atomic bomb could have been ready a full year earlier. The Nobel laureate physicist Isidor I. Rabi accused Szilard that his actions saved the Germans from an early American atomic bomb.[112] (The feasibility of

---

[112] Ibid., p. x.

offering Szilard employment would come up in the Second Conversation).

When Watson refers to a Szilard idea, it makes direct sense too because the Hungarian word "szilárd" means solid in English.[113]

Money is often another motivation. Today even graduate students in molecular biology think about selling their science.

I was never brought up with money as a value; we were poor, we didn't have a car; it was ideas that were the great thing.

What did your father do?

He collected money and not very successfully. Someone owed a bill and he was trying to collect it.

Not a very pleasant occupation.

No, it was terrible.

Your mother?

She worked with personnel at the Red Cross during the war. We lived in Chicago. Then she worked in the office of admissions of the University of Chicago. She wasn't an intellectual but my father was. She was a people person; my father liked ideas. They were slightly misfit but both were very nice people. For him ideas were the main thing and in that sense I had a Jewish upbringing.

Did you realize this at that time?

Then I knew only that Jewish food was awful.

Considering the Jewish component in science today many people find hard to imagine the extent of discrimination against Jews in higher education and in the job market before the war in the United States.

Sure. This is why Luria and Muller were in Indiana although I'm not sure about Muller but people assumed he was also

---

[113] Szilard was born Leó Spitz and the family changed their surname to Szilárd when he was two years old. When he left Hungary the "ó" and the "á" in his name gave way to "o" and "a." The "ó" and "á" in Szilard's original name were not expressions of indication of accents as many non-Hungarian authors assume; rather, they signify different sounds from "o" and "a."

Jewish. There were just some Jews who were not religious and there were others who wanted to assimilate. The Jewish religion is tough. I have a lot of Christian values because I was raised that way.

What would be your Christian values that are different from Jewish values?

More help to the underdog, "Blessed are the weak." I'm terrible on religion but when I was in Jerusalem I did some reading and I thought it was just too tough, the "eye for an eye," all that. The whole tradition is a bit overbearing. We just saw a Woody Allen film. Christianity is lighter. I like to be light although I know the world isn't. In the BBC movie about the double helix I was played by a Jewish actor, Jeff Goldblum. I thought this was unfair because when I grew up all scientists were Jewish, all theoretical physicists and all biochemists were Jewish, with a few exceptions. Eighty percent of the students were Jewish in science. Now, not even 2%, it's terrible. I should've been portrayed as Irish, the underdog; this is what my Christian sense would suggest; the underdog who always succeeds.

While Watson correctly assesses the presence of Jewish professors at Indiana University, he got it wrong about the universal Jewish presence in science at the time of his growing up. He was 15 years old when he finished high school and that was 1943. Isidor I. Rabi was the first Jewish physics professor at Columbia University (1937),[114] Charles Yanofsky[115] was the second Jewish professor of biology at Stanford University, and Carl Djerassi was the first Jewish chemistry professor at Stanford University, which he joined in 1959. According to Djerassi, this was "not because Stanford was anti-Semitic, not at all, it was just not yet done."[116] Both Herbert Hauptman and Jerome Karle, the 1985

[114] *Nobel Lectures in Physics 1942–1962*. World Scientific, Singapore, 1998, p. 20.
[115] Conversation with Charles Yanofsky at Stanford University on September 8, 2006, unpublished records.
[116] Hargittai, *Candid Science*, p. 87. (Carl Djerassi)

Nobel laureates in chemistry for introducing the direct methods in
X-ray crystallography, had difficulties in getting into graduate school.

Karle graduated from City College of New York in 1937 when he was
19 years old. What happened then is transmitted here in his words:[117]

> "It was virtually impossible to get into medical school for rea-
> sons that are not very complimentary for the society at the time.
>
> *Were there actual rules against Jews or it just happened?*
>
> There were no rules. I don't even know whether I should
> speak about this.
>
> *Please, do.*
>
> I went to Harvard and spent a year there. I had this stupid
> illusion that being a good student was all that was necessary
> to get admitted. I applied to Harvard and some other places
> and, of course, I was turned down. I wanted to try again and
> I was allowed to have a conversation with the Dean. The only
> thing I got from him was a harangue. He said, 'We have
> enough Jews in Massachusetts we don't need any from New
> York City.'
>
> *What did you do then?*
>
> I had applied to various graduate schools just to do graduate
> work and I was turned down by all of them. So I wasn't doing
> anything. Then there was just a stroke of luck. In the summer
> of 1938 I was working on Coney Island and a good friend of
> mine with whom I still communicate told me that they were
> having exams for a civil service job in the New York State
> Health Department. I took the exam and I had the highest
> grade among those they accepted. They simply had to accept
> me. Then there was a rule that after a certain period, maybe
> three months, the Health Department could not dismiss any-
> body without an explanation. I stayed for two years. I learned
> only later that they had wanted me to leave but my boss said

---

[117] Hargittai, *Candid Science VI*, pp. 422–437. (Jerome Karle)

that if they tried to dismiss me, he would make such a fuss that I was kept. During those two years I was saving up money as I knew that I couldn't get any money from graduate schools. At that time someone told me that if I went to the University of Michigan, there I would be treated properly. That's why I went to the University of Michigan."

So Karle went to Ann Arbor and became a physical chemist. For his doctorate, he chose gas-phase electron diffraction and molecular structure determination. Lawrence Brockway, a former student of Linus Pauling was his supervisor. Although Karle finished at the top of his class, he was not given a teaching assistantship. When Brockway protested to the dean, who was a German and had studied under Moses Gomberg, a Jewish professor and discoverer of free radicals, the dean told Brockway that he never gave teaching assistantships to "Jews, Negroes, Italians, and women." Karle never found out what Brockway told the dean, but he had his teaching assistantship by the next morning. This happened in 1940.[118]

Let here also be a minor comment on Watson's statement quoted above that "Jewish food is awful." Apart from the fact that I do not share his opinion, I suspect that this is one of the *watsonisms* that he likes to apply for a minor shock value. Max Perutz wrote Watson a farewell letter on January 14, 2002, three weeks before he died in which he quoted some of Watson's statements from their first meetings. When Watson did not like to be in Jesus College, Perutz helped him transfer to Clare. When then Perutz asked him if he was happy there, and Perutz expected some words of appreciation, all what Watson replied was "The food is awful!"[119]   That

You were the underdog?

I was because I wasn't brought up to be so tough.

[118] This story is described in Hargittai, I., *Our Lives: Encounters of a Scientist*. Akadémiai Kiadó, Budapest, 2004, pp. 73–74, based on a conversation with Jerome Karle in Washington, DC, 2000.
[119] *Inspiring Science*, p. 73.

Who is the underdog today?

The European; because they are not moving so fast. I have European culture. To me London is the greatest city in the world. You just walk around and you see history. If you think of great civilizations, it's the British and the Russian. If you ask me who are the best people around here, they are the Russian Jews. They're interesting and alive; you can compare them to the Hungarian Jews.

This was the end of our first conversation.

We had hardly returned to Budapest when I received a letter from Watson dated May 22, 2000, saying that "I apologize for not being able to better host you during your visit on Saturday. I am still much too heavily programmed. On your next visit to the States, you must come again for a true visit to our village for science." Then, soon, the Watsons visited us in Budapest for one day in July. We had only a few hours for sightseeing. The open-air sculpture garden where the sculptures of the communist era have been gathered was scrapped because of lack of time. We visited the Holocaust Memorial and the Minta high school mentioned before. Liz was interested in sights of architectural importance in keeping with her being an architectural historian. She also had a special request to visit the Main Synagogue, which is the largest in Europe or at least in Central

Jim Watson taking photos; in the Hargittais' home; and with Liz Watson at the City Park, in Budapest, 2000 (photos by the author).

Europe about which she had heard from a friend. We visited the Heroes' Square with the Defending Angel of Hungary standing on top of a tall column. In front of this column are the seven leaders of the Magyars who led them into the territory of present-day Hungary. Behind, in a huge semi-colonnade stand the statues of the most important kings and sovereigns of Hungarian history. Further back a small bridge crosses an artificial lake, leading to an ensemble of buildings built at the end of the 19th century, replicas of various architectural highlights from all over Hungary. The Watsons appeared genuinely interested in what they saw and heard. During our walks Jim was always charging ahead, Liz was a more contemplating sightseer, sometimes disappearing at corners or in shops, turning up with yet another guidebook.

When the visit to the City Park was over, Jim suggested going into the Museum of Contemporary Art because he noticed there a temporary exhibition of Klee, which he was keen to see because they have two Klees at home. Leaving the museum, Jim remarked that Hungarian museums are lighter and more spacious than the Austrian ones. This was one of similar generalizing remarks he made. My favorite Watson generalization came after we had had ice cream. In the city center, we saw several large cafes, restaurants, and ice cream parlors, but we decided to go for ice cream in a side street to my favorite small parlor. Jim twice exclaimed during eating the ice cream that it was unbelievably good. Jim asked me whether people here would eat more often ice cream in the morning or in the afternoon." Later Jim repeatedly referred to the Hungarian habit of 5 o'clock ice cream. I admired this little episode — perhaps reading too much into it — but it seemed to me as if observing Jim's mind working in the miniature. Before making any remark he collected as many facts as was practicable under the circumstances. In this case, for example, he sampled the ice cream and he inquired about the ice cream eating habits of the Hungarians. Then he made his conclusion based on rather limited amount of information. He also packaged the conclusion in an easy to perceive and understandable format.

Jim Watson with the author and with Magdi Hargittai in the Hargittais' home in Budapest, 2000 (photos by M. Hargittai and by the author, respectively).

Jonas Salk and Albert Sabin on US stamps (2006). According to Watson, they should be canonized.

When we corresponded about the visit, it was clear that Jim did not want to make laboratory visits. Rather, he mentioned that he would like to meet with Hungarian intellectuals because that is what Budapest is famous for. We took the request seriously and gave much thought to the guest list for the evening party in our home. George and Eva Klein of Stockholm knew about our party and telephoned us the next day asking about how it went. Müller-Hill happened to be visiting with them from Cologne and they told us that they were playing a game — Who would be the dozen Swedish intellectuals and the dozen German intellectuals that they would invite for a similar party in Stockholm and in Cologne?

One of the most remarkable comments Jim made during the party was to the question about the biggest benefit from the completion of the Human Genome Project. He said it would show the average Hungarian the validity of evolution versus creation. Jim told us about his forthcoming visit to the Vatican. He had plans to suggest making Albert Sabin and Jonas Salk saints. Jim thinks that the Catholic Church is going to disintegrate unless it renews itself.

He enjoys controversial topics, and is amused by the interest he generated with his remark that fat people enjoy more their sex life than thin people. One also wonders, where would have he collected his information about fat people's sex lives. It seemed to be a calculated controversy because Jim seems to be very careful in choosing which topic to appear careless about.

After the Watsons' visit I told George Klein about how appreciatively Jim talked about him. George was surprised because they had no interaction after their exchange of letters in 1967. So George sent Jim one of his books, Jim reciprocated, and their interaction resumed, as a side benefit from the Watsons' Budapest visit. Their break in their interaction between 1967 and 2002 must have followed from some unclear misunderstanding. George was one of the many to whom Jim had sent his manuscript of what later became *The Double Helix*; at that time it was yet *Honest Jim*. George responded with a most positive and encouraging review, which was significant because many of the preliminary reactions to the manuscript were negative.

The question of Jewishness came up at different points in our Budapest conversations. Watson is intrigued and fascinated by this topic. Above there was already a mention of his exaggerated projection of Jewish presence in science at the time when he was growing up. As if to compensate his fascination with and admiration of Jewish

Jim Watson looking through the fence at the Holocaust Memorial in the yard of the
Main Synagogue in Budapest, 2000 (photo by the author).

intellectual achievements, Watson makes negative comments about
other aspects of Jewish life. As we have seen, he likes to state that he
hates Jewish food. He also declares Jewish religion too tough
although he admits that he knows next to nothing about it. When he
stressed that, in spite of being non-religious, he has a lot of Christian
values, this intrigued me. This is often emphasized as if being in contrast to Jewish values. Former US Secretary of State Madeline
Albright had made such statements after she had learned — what was,
ostensibly, news to her — about her Jewish roots. I had no opportunity to ask her what she meant by that, but I could ask Jim about it.
We have seen his response above, but what the printed version does
not reveal is that he had to think long and hard before he came up
with an answer because the question found him unprepared. I guess
people like to make such a statement, which usually ends that part of
the conversation rather than getting a follow-up question.

At the same time, I find it hard to remember anyone who would have spoken so disparagingly about religion and the Catholic Church in particular, as Jim. To me one of the highlights of our conversation was when Watson spoke about the pre-eminence of knowledge over prayer in solving the world's problems.

Second Conversation

We want truth to come
from observations and experiments.

James D. Watson

In our exchanges I told Jim about the book project I was working on, which he liked and suggested that I come to Cold Spring Harbor for a few months to work on the book. The project was in part autobiographical.[120] I never kept a diary, and hardly ever preserved old documents, which may have been in line with our family having never had an archive because our records got lost during voluntary and forced moving. Around 1990, however, my perception about the past altered. The political changes brought a democratization of our society, and one of the by-products was overt anti-Semitism. This anti-Semitism must have been around during socialism but it was latent, suppressed, and now it burst out. This new phenomenon made me think about our past, the killing of my father in a forced labor camp, our deportation to a concentration camp in 1944, the circumstances of our return and the start of our new life in 1945. Then, in the 1950s, I had difficulties in getting first into secondary school and four years later into university. The pretext was my bourgeois origin based

---

[120] Hargittai, *Our Lives.*

first on my late maternal grandfather's having been a shop-owner before World War II, and later it was based on my stepfather having been a shop-owner for a few years after World War II. I decided to record these stories and others lest they be forgotten. In terms of childhood experience, I was probably Jim's opposite who could claim having been an underdog, because he was not brought up to be tough. I could have qualified for having been brought up to be pre-pared for the worst. Yet I used to consider myself lucky. I survived the inferno of deportation and the camp. In spite of all hurdles, I could go to all the schools I had wanted to. Finally, I could practice a pro-fession — scientific research — that I had always dreamed of. I did not want a whole book devoted to me, so I placed my story among my encounters with famous scientists. The book appeared two years later and was well received.[121]

We arrived in Cold Spring Harbor in January 2002 on a snowy day. Jim had organized our transportation from Kennedy Airport to CSHL in an elegant limo. Liz was waiting for us at Williams House, which was to be our home for the next three months. She brought flowers and various little things, like pottery, and a bottle of wine to make us feel more at home. She also noticed the need for some minor repair and the next day there was a flurry of activity by various work-men in our home. Jim was aware of everything and checked person-ally their progress during the next few days. I found the span of his attention hard to believe. The evening of our arrival the Watsons had us for dinner; obviously Liz did the cooking, and Jim was exercising the duties of an impeccable host. This attention continued for the next three months whenever they were in residence and they were not always. Jim's obligations kept them away at times, and at other times they had to be staying in New York City to be at hand as their older son was in the hospital.

During our stay, I wanted to record a second conversation with Jim. The first conversation left good memories with me, but a sense of incompleteness, too. For the second recording, once again a "non-work-day" was appointed, January 21, 2002, Martin Luther King Day. We met

---

[121] See, e.g., *Nature* 2004, **432**, November 11, p. 150 (review of *Our Lives*).

Jim Watson and the author at CSHL in 2000 (left) and 2002 (photos by M. Hargittai).

in his office and the second recording was, again, planned for one hour. This time I did not feel the tension that I had to make this conversation happen and I did not have a set of questions because by then I had had several opportunities to talk with Jim. The only difference from our other chats was that now we had the Dictaphone running.

There was a most significant event between the first and second recordings. Less than half a year before, the terror attack of September 11, 2001, happened.

What sets scientists apart from the rest?

It's basically that we want truth to come from observations and experiments. Many people think that truth can come from revelation. That's a big difference. We don't want to believe in God because everyone else does. There has to be reason.

Today hundreds of thousands of people work in science. We can no longer consider scientists a select and creative group of people.

Many of the people, probably a vast mob do not probably reject revelation. A million of people work with DNA and I'm sure a lot of them go to church on Sunday. But if you ask about how many people attend services during the week who are members of the National Academy, the number is very small.

I would distinguish two groups. One goes there because it's a social habit and the other group does because it truly believes. In writing my book, I am not concerned with religion. To me, my Jewishness has nothing to do with religion.

That's the trouble with Jews now. They feel now that they have to identify with religion. When I grew up the Jews I knew were all reformed. Now they feel the necessity to ascertain their Jewishness in a way that the people I knew did not feel the necessity for. Fifty years after the Holocaust people are using it as a justification for Bar Mitzvah and other archaic crap. I find it hard because to me a multicultural country is oxymoron. You have to have one culture. When you fight a war you have to believe in something. You have to have a unity for your country. Because of the awfulness of the Moslem religion, the American Moslems have not come out against the terrorists. The only solution for an effective country is hetero-marriage. The Irish problem disappeared when they started marrying other people. Religion divides people. There is no future in being a Jew in the United States. There is only a future in being American. There is no future in being a Catholic; I just find the whole thing a mess.

Don't you think then that it is hopeless?

No. If you discount the Hassidic, half the Jews marry non-Jews. The Catholics marry Italians.

Italians are usually Catholics.

Now the whole thing is vanishing.

How about Northern Ireland?

The only future there is to do away with the two sets of schools. As long as you have the two sets of schools, it's hopeless. In any case we are preaching that we can maintain our individualities and still have a national one. One has to be more important than the other. The Moslems are afraid to speak because they get a Jihad against them. You can be an ex-Catholic but you can't be an ex-Moslem without them wanting to kill you. It's awful.

In Hungary there is strong anti-Semitism. I was wondering whether a recipe like the civil rights movement in the 1960s in the United States would work to eliminate it.

No. I don't think there is any solution except intermarriage.

I remember Lyndon Johnson's speech to very wealthy people; I think it was in Houston in 1964. He said that there are all kinds of problems, but there are people who can't do anything else but shout, "nigra, nigra, nigra."

The only solution to the black problem is mixture.

But there has been tremendous progress even without mixture.

I don't think that you can have a minority within a country that sticks to itself; intermarry, and be totally accepted by the majority. It's asking too much. In Switzerland, in some complicated way, with the cantons, which control their own fate, they've managed.

In Switzerland, I have visited mostly the German part. On one occasion I was invited to give a series of lectures in the French universities. As I was leaving Bern and German Switzerland, my friends were terrified and when I asked them what their problem was, they asked me back whether I have not heard that in the French canton people were Catholics. I used to think that Switzerland was a country free from such considerations.

The only solution is intermarriage. The Indians first marry Indians but eventually they break down after two generations. The Pakistanis in England probably should just go home. They are being hated and they should be hated because they were all for the World Trade Center bombing. Just get rid of them. What I am trying to say is that unless you adapt to the values of the country ... Of course you can say that the country has lousy values, then you got to leave the country.

You could not run for office on such a platform.

I suspect that multiculturalism has reached its peak because it's crap. As long as there is an Afro-American department, blacks will be discriminated against.

The experience of anti-Semitism may encourage some to hide, but for many, it strengthens Jewish identity.

Before the Holocaust, the biggest Jewish banker in the United States was Otto Kaahn, who was probably born in Germany. He and J. P. Morgan were the two big bankers, one Jewish and one WASP. Kaahn is buried in the Protestant cemetery. He saw no future for himself as a Jew. He wanted to be American.

And Protestant.

That was the socially proper thing to do. He had the Duke of Windsor in his house. One of his sons married a Protestant woman, Pauli Kaahn. We've been good friends. It was a group, which basically said that we're going to merge. It's rooted in the human nature, this tribal business, fear of other groups whose loyalties are to the groups rather than to the country. It's a hard thing but I am cynical about solutions. The Protestant/Catholic division in the United States, except some parts in the South, is vanishing because of intermarriage.

Would you suggest then to the Jews who would not go along with such a plan to go to Israel?

If they really want to have Jewish laws and be considered primarily as Jews and not Americans, yes, go to Israel. If your loyalty is to something else, and in many cases it is. I instinctively want to get away from people who are too Jewish. I find them very limited and boring. Everything has its time. We have so many problems in this world that I would like religion to cease being the cause of so many problems. When I see a Moslem, I just want to get away from him. I was reading the *Moslem Brotherhood*. They believe that the whole world should be Moslem. They state it. They'll only be satisfied with everyone being a Moslem. The people in Israel are totally unrealistic. They think they can take away land, and there should be no terrorism. If we take away people's land, there is going to be terrorism. We take away the Palestinian land and they don't have the right to shoot us? I'd shoot them. If I was a Palestinian I'd be a terrorist. I think Israel and the Jewish lobby is not reality. Most people don't say this because they do not want to upset the Jewish community and they don't want to be called anti-Semite. I am not anti-Semite, I just find it intolerable. That's why I didn't go there for 40 years. Then I went there 18 months ago because it looked like there might be peace and the settlements would stop. Barak was trying to but then there was that idiot Arafat.

I was having very high hopes too.

Arafat is just another Moslem who could not co-exist with other people.

He is also a shrewd politician. He speaks one thing in English and something else in Arabic.

I don't think anyone believes him now, his credibility is gone. The sooner he is dead the better. Get Arafat out. The Moslem world is so wretched and it's getting worse. It can't compete with the West given its values.

We started from what is different for scientists.

I once gave the commencement address at the University of Judaism at Bellvere. It's conservative with the first conservative rabbis outside the East Coast. It was a most boring place. I would never want to go to dinner with these people. They're boring whereas the Jewish world is the most exciting...

Outside this confinement.

Outside this confinement.

Michael Polanyi gave a speech in England in 1943, which was a very difficult time. He noted that as the Jews were leaving the confinement of their community they became very exciting people but not within the confinement.

This is because of the Jewish cultural tradition. I now tell people that I'm culturally Jewish, which is why I can move in a lot of different worlds. I'm not held back. Just tell me what the truth is; this is part of the Jewish thing. But when it is put into the orthodoxy, oh God, you don't see the truth.

At the Nobel Centennial, Magdi and I were sitting opposite to you and the other Nobel laureates in the Concert Hall during the award ceremony. We took many pictures, but Magdi took pictures also segment by segment of small groups of the laureates because her camera had an especially strong zoom. In one of her pictures you are seen in the center with the other laureates around you. You look as if you were out of place in the sense that you are looking around as if looking for something.

I was bored to death.

How can you find a more intellectual company?

Oh, they're just too old. Except for the younger people the only person who was truly alive was Steve Weinberg.[122] He was still tough. It's hard to be tough all your life and the others ceased being tough.

---

[122] Steven Weinberg is at the University of Texas at Austin where he moved in 1982 from Harvard University. He shared the physics Nobel Prize in 1979 with Sheldon L. Glashow and Abdus Salam "for their contributions to the theory of the unified weak and electromagnetic interaction between elementary particles, including inter alia the prediction of the weak neutral current." See, also, Hargittai, M., Hargittai, I. *Candid Science IV: Conversations with Famous Physicists*. Imperial College Press, London, 2004, pp. 20–31. (Steven Weinberg)

In the center, Watson, surrounded by other Nobel laureates at the Nobel Prize Centennial in Stockholm, 2001 (photo by M. Hargittai).

I was glad that there was no declaration by the Nobel laureates about anything.

[John] Polanyi[123] wanted to do something and I thought it was crap. Then there was another one, Ferid Murad,[124] who is down in Texas, the nitric oxide guy. He wanted to have a petition.

---

[123] John C. Polanyi is at the University of Toronto. His father was Michael Polanyi, originally a medical doctor turned physical chemist who later in his life switched to epistemology. Although John Polanyi is a Nobel laureate (Chemistry 1986), he has to live in the shadow of a great scientist father. He seldom misses to remind his partners in conversation that being the only living Canadian Nobel laureate, he has special obligations. See, also, Hargittai, *Candid Science III*, pp. 378–391.

[124] Ferid Murad received his share of the Nobel Prize with Robert F. Furchgott (see, Hargittai, *Candid Science II*, pp. 578–593) and Louis J. Ignarro "for their discoveries concerning nitric oxide as a signaling molecule in the cardiovascular system." In addition to positions of researcher and educator, Murad has often occupied company positions during his career. There was a big controversy about the composition of awardees of this Nobel Prize. Furchgott was an unambiguous choice for having pioneered the field. However, many protested the absence of Salvador Moncada (see, Hargittai, *Candid Science II*, pp. 564–577) from among the awardees. Furchgott himself stated that Moncada should have been included, but he would not elaborate on who might have been displaced (Hargittai, *Candid Science II*, p. 593).

About what?

It was about helping the poor, helping the Third World, implying that America is not perfect. Polanyi's, I thought, was particularly awful. It was as if the World Trade Center was America's fault.[125] That we're not worrying about the environment, we're not worrying about global warming, that Nobel Prize winners should lead the world. I don't think Nobel Prize winners could lead anything.

As we were being taken in a bus from the Concert Hall to the City Hall, to the banquet, I wonder if you noticed that there were some protest groups.

I didn't.

You may have been sitting in the other side of the bus. Their poster said, "Free political prisoners in Iran."

I met there a Swedish scientist, active in a committee of the Royal Swedish Academy of Sciences that is concerned with the political freedom of intellectuals or something like that. His wife, who is American, came up to me at the Royal Palace and attacked me for wearing an American flag, saying that America was a corrupt society. She is a very, very rich, spoiled leftist person. I told her why doesn't she sell her diamond necklace and give it to the oppressed. She said it was her mother's. She's just a rich bitch. I just realized why they cut the heads off of very rich people. That's the only thing you can do.

I talked with Henry Taube,[126]

He is a nice man.

---

[125] The Nobel Prize award ceremony took place on December 10, 2001, that is, a mere three months after the terror attacks of September 11, 2001.

[126] Henry Taube was Professor of Chemistry at Stanford University. He received the chemistry Nobel Prize in 1983 "for his work on the mechanisms of electron transfer reactions, especially in metal complexes." He was born in Canada, but his parents immigrated there from czarist Russia. They were almost illiterate peasants whose German forebears had been imported to Russia during Empress Catherine's reign in order to improve farming practices among Russian muzhiks. See, also, Hargittai, *Candid Science III*, pp. 400–413.

and he told me that when we are happy about the demise of the Soviet Union, and rightly so,

Yes.

we should not forget that the Russian revolution came because of this terribly oppressive czarist regime.

Oh, yes.

We tend to forget this.

Let me ask you something. We were in Sankt Petersburg for two days. There was a big party paid for by Paul Allen of Microsoft. He is worth US$30 billion. There were 200 of us on a cruise ship. The first night we went to one of the big Sankt Petersburg palaces, which had been taken over by the trade unions or something. This was a place where they had ethnic evenings. You could see clowns, circus performers, the traditional Russian things. Watching the clowns I suddenly realized that I was looking at Marx brothers. Then I wondered to what extent is what we call Jewish humor and Jewish culture Russian?

To a great extent. Israel was founded mostly by Jews from Russia, which included Poland at the time.

Even though there was vast anti-Semitism in Russia, the culture is really a Russian culture.

Magdi and I, when we meet Russian Jews we first identify them as Russians and only then as Jews because it is undeniable that they are Russians.

That's what I'm trying to see. Because of the oppression, the Russians were just like Jews.

It is rumored that some Russians immigrated to Israel declaring themselves Jewish although they were not, and became Israelis.

OK, so the Russian humor forced itself into it.

It's very difficult to distinguish which part of Russian culture is Jewish in origin and vice versa. It's inseparable.

That's what I thought and since there are more Russians than Jews ...

The Jews were terribly active and as soon as the oppression eased ... Of course, humor thrives when there is political oppression.

So Russia had more oppression.

The recent Russian immigration brought to Israel more Russian customs than just humor.

There were no paintings in Russia until Peter the Great, just icons. Then suddenly there was 200 years of very great vitality until the Revolution. It's a great culture.

After the Revolution, in the 1920s, Russian painting was pioneering.

Anyway, it was very interesting just seeing the clowns and everything else and where it comes from.

The former West German Chancellor Helmut Schmidt had said long before the Berlin Wall had come down that the Soviet Union was 25% communist and 75% Russian.

Hitler would've eradicated everything.

Both the communists and the Russians

and the intellectuals. Today it is religiously important to be politically correct. When I went to the University of Chicago, it was extremely liberating for me, at the age of 15. There were people who were actually trying to use ideas. It was really ideas.

One of my distant heroes is Robert Hutchins. You and others have told me about his innovations, that he compiled a list of books. He found them more important than special topics. He did not mind if bright young people sailed through the university in two years, and he gave a job to Leo Szilard.

Sure.

The job was rather loosely defined.

Because no department would take him. He was Professor of Social Sciences and Biophysics. The University of Chicago stood for ideas.

You were lucky.

I was very lucky.

Speaking about intellectuals, did you see this book review about public intellectuals in The New York Times?

Posner is the author.[127] I went to the bookstore but it wasn't there. It's probably a slightly phony book. It seems to be a citation index.

I looked up myself on the Internet and have appeared 700 times. I looked up Francis Crick, 30,000. I looked you up, and guess your number.

Fifteen thousand.

Seven hundred thousand. [These were end of 2001 data.]

Oh. OK, that's probably because Francis has never been a public figure and I have been.

Tremendous difference.

I think it's rightly because I'm on committees. Francis never wanted to be on any committee in his life. Have I told you about Francis's only public statement in his life?

No.

When the Rolling Stones were busted, he signed a petition to legalize marijuana. That was Francis's only time.

Last time you told me about his speech at the University College London.

Yes.

---

[127] Posner, R. A., *Public Intellectuals: A Study of Decline*. Harvard University Press, Cambridge, Massachusetts, 2002.

*I wrote to him afterwards and he wrote me back that yes,
he remembered the speech and he would give it slightly different today. By the way, he authorized me to tell about what he
wrote me.*

Sure.

Watson's critical remarks with respect to politics and Israel are not dis-
similar to some others in the West, who view the plight of Israel's
struggle for survival from the comfort and security of the great
democracies. Many find it expedient to distinguish between the acts
of terror against Israel and against the rest of the world. It seems as if
Watson's sympathy for Jewish respect for science and education
would prod him to be more critical toward other aspects related to
Jews and Israel from the trivial to the vitally important.

Watson's reference to Crick's lecture in London in 1968 in our
first conversation made me ask Francis Crick about the two questions
referred to by Watson. One was Crick's suggestion to count life as
starting after the first two days of the baby's life, and the other was
not to spend money for medical care on people above 80. Here is the
answer from Crick dated June 28, 2001:[128]

"I did indeed give a provocative lecture in 1968 (or there-
abouts) at University College London, but I'm not sure that
I still have a copy of it.

To reply to your two questions I would indeed modify my
suggestions today. In the old days doctors quickly let a very
deformed or handicapped baby die, rather than make exceptional efforts, as they often do now, to keep the baby alive.
I now realize that it would be impossible, at least in this
country, to count life as starting after the first two days of
the baby's life because so many religious people believe life

[128] Francis Crick's letter of June 28, 2001, to me was first printed as a footnote to my first conversation with Watson, Hargittai, *Candid Science II*, p. 10, and later as part of a larger account
of my correspondence with Francis Crick and our visit to the Cricks, Hargittai, *Candid Science
VI*, pp. 2–19.

Francis Crick in the Cricks' home, La Jolla, 2004 (photo by the author).

effectively starts much earlier, even at conception. In other words one has to consider not just the feelings of the baby (who hardly has any) but also the feelings of the parents, and of other members of society, however silly one may think them to be. But I do believe that doctors should not make exceptional efforts to keep a very handicapped baby alive.

As to the age limit, people now live longer than they did in the sixties, so I think such an age might be a little higher, but I doubt if a rigid rule would be acceptable. Again I think very expensive treatments, or ones that have only a limited availability, should be allocated in some sensible way. I've heard that the State of Oregon is trying out such a scheme.

If I were to give such a lecture again (which is unlikely) I would instead stress the right of a person who is incurably ill to terminate his life. I believe this is being tried out in Holland."

Back to the Second Conversation:

Crick has slightly modified his views; he has made them milder, especially because he has realized that public sentiment would never tolerate his radical approach.

My belief is that that's a failure of Francis because someone actually has to be out there and saying it. Then the middle will say this, look, Francis says this, and you could move slightly toward it. If you move toward the center, you are not serving your function as an intellectual. If you're a politician, you can't say it, but Francis can. I'm heavily restrained because I'm head of an institution and we have to raise money.

He did not say that he changed his mind, only that he realized that it was impossible.

I think that in a hundred years from now it may not be, because people will realize the horror of raising a child with bad cerebral palsy. It's torture, unbelievable, and there is nothing one can do. You should not allow when you see a child with terrible cerebral palsy to live.

A friend of ours had a baby with a minor birth defect. His wife killed the baby. She is now in prison. She had been taken to a hospital for medical observation because there were strong signs that she was having problems, but then they let her go home. Then the tragedy came, which ruined the life of several people. When this happened, I remembered our previous conversation about this problem.
Sure.

In my letter of August 8, 2003, to Crick, I mentioned that Gunther Stent told me[129] about an essay in 1973 in which Stent presented a

---

[129] Hargittai, *Candid Science V*, pp. 480–527. (Gunther S. Stent)

linguistic analysis of some of Crick's writings. Stent substituted the word "God" wherever Crick had used the word "nature," and — according to Stent — the substitution did not change the essential meaning of the text. I also asked Crick for comment in connection with the Church's recent acceptance of the idea of evolution. I mentioned Wigner's position that physics did not endeavor to explain nature, it only endeavored to explain the regularities in the behavior of objects. In his response dated August 29, Crick responded to this as follows:[130]

"Gunther Stent, as usual, has produced an entertaining mixture of sense and nonsense. I will only say that my position is that I am an agnostic, with strong inclination towards atheism. For Gunther's term 'nature' I would prefer 'The Entire Universe.' I agree with Wigner that a little modesty would not be out of place.

I will not comment on the so-called 'religious' views of Einstein and Bohr. Gunther's remarks about Babylon, etc. miss the point, which is that Darwin effectively discredited 'The Argument from Design' which before him seemed unanswerable. Enough of my old friend Gunther!

By the 'Church' I presume you meant the Catholic Church. All the 'religions of the book' (the Bible) differ substantially among themselves. All three have both extensive sects and sub-sects. A recent encyclical by the present Pope said evolution must now be regarded as a fact, though it disapproves of what I do now. But in the USA millions of, say, Southern Baptists think evolution is quite wrong, that the earth is less than 10,000 years old, etc."

In our conversation with Crick in 2004, his views on religion came up. I mentioned to him that Watson was not happy that Crick seemed to have been moving from a more radical position towards the center. According to Watson, Crick was in a better situation to

---

[130] Hargittai, *Candid Science VI*, pp. 2–19. (Francis Crick)

criticize religion than Watson was because Crick was not the head of a major organization and he was not involved in fund-raising as he, Watson, was. Crick said that to fight religion at the present time produces only frustration. First we have to understand how the brain operates and after that it will be much easier to convince people that religion is meaningless. We also talked about the recent changes in the views of the Catholic Church regarding evolution, for example. Crick stressed, however, that the Catholic Church seems to want to solve all the problems of religion within its own framework and without the involvement of science.

> *Coming back to Francis Crick, I exchanged letters with him about something else too. Most of my interviewees that named their heroes, named him as one of them.*
> Sure.
>
> *He appears very influential among this very select group.*
> Yes, sure.
>
> *So I asked him about pupils and his impact on them from his point of view. He said that he could not work with people, except for a few partners, a total of about three. You were the first.*
> Yes.
>
> *Then Sydney Brenner.*
> Yes, and Christof Koch. He is a neuroscientist at Caltech.
>
> *Crick was very good in one to one relationships. You had students.*
> A lot. Feynman never had a student. I don't think anyone was ever a student of Feynman. I think it was because he wanted someone who could challenge him and at the level of the students they couldn't challenge him, so he was not interested in that. He gave very good lectures, but he probably didn't want to give a course.

In our correspondence, I asked Crick (on July 27, 2001) whether there were any scientists that could be considered directly as his pupils. Again, Crick's response of August 1, 2001, is of interest in full:[131]

"To reply to your question, I don't think there is anyone whom I could call my pupil. I only supervised a graduate student for a year, but after that year someone else took him over. I think I deliberately avoided such tasks.

On the other hand I have had several close collaborators. The major ones have been Jim Watson, Sydney Brenner and (more recently) Christof Koch. Others I have had more than transient collaborations with are Aaron Klug, Beatrice Magdoff, Leslie Orgel and Graeme Mitchison. In all these collaborations we have published papers together. These collaborators (except possibly for Magdoff and Mitchison) have each had many pupils of their own.

I think I work best, not entirely by myself, but with one other close collaborator. Sydney Brenner and I shared an office for 20 years. At the moment my close collaborator is Christof Koch, a neuroscientist at Caltech.

Of course I have interacted for most of my scientific life with a very large number of scientists and over the years have given lectures in many different places. Some people have told me that they were strongly influenced by a lecture of mine they heard. I think I must have been a rather good lecturer, because at meetings no one liked to have to lecture after me!"

Speaking about Crick's lectures, Frederick Sanger told me that at one time he tried to emulate Crick's style of lecturing. He tried to make his lectures more colorful. But the jokes did not work and he gave it up. Sanger is a very modest person.

Of course, and is boy's honest, as they come. Sanger was enjoying himself in Stockholm. He was trying to be more than he was. He is such a nice person.

---

[131] Ibid.

We visited Sanger last year in his home. LMB seems to neglect him.

Sydney [Brenner] fired him. Sydney walked through his lab and showed someone his space that they were going to have. Sydney wanted to get rid of him.[132] He was jealous. Sydney is not a nice person.

Wasn't it rather futile to be jealous of a person who has won two Nobel Prizes?

But Sydney has zeal.

When we visited the Cricks, we told them about how much Sanger enjoyed Crick's lectures and his style of lecturing. He would have liked to emulate Crick's ease of lecturing and the way he inserted his jokes into his lectures. At one point Sanger had decided to give it a try and had carefully prepared his presentation in Crick's style, but he came away disappointed because his jokes came out flat and were received in silence. Crick enjoyed the story and he demonstrated his famous roaring laughter. I find it moving when I remember that this was a few months before his death.

Crick said that the jokes did not work the way Sanger imagined them. He never prepared his jokes specifically although they came out of his reservoir; but they came out — or appeared to come out — spontaneously during the talk. Once he was asked to give a lecture in Paris, but he was to give it in French. His French was not that good, so he wrote it up and his wife, Odile corrected and translated it. The first thing that had to be left out was the jokes; it would have been difficult to have his old jokes in French, besides, planning them in advance — so that they could be translated — made it impossible for them to appear spontaneous.

When asked who his heroes were, Brenner told me:[133] "Francis [Crick] is a hero for me, and Fred [Sanger] is another kind of hero.

---

[132] Aaron Klug tells the story more accurately: "When Sanger retired at the MRC age limit, Sanger asked if he could stay on in his old Division of which he had been Head. Sydney said he couldn't do that, but could stay on if he made arrangements with another division." From Aaron Klug's letter of September 26, 2006, to the author.

[133] Hargittai, *Candid Science VI*, pp. 2–19. (Francis Crick)

Odile and Francis Crick during our visit, La Jolla, 2004 (photo by the author).

Fred is a hero for just going about doing things. He is a craftsman of science. Francis has a more brilliant mind, but Fred has the craft of doing things. He is a good chemist; that's what Fred is, a superb chemist."

Sidney Altman told a story[134] that illustrates Sanger's modesty. Long before his Nobel Prize, Altman was doing his postdoctoral research at the MRC Laboratory of Molecular Biology. In an experiment, the bubbler tube was punctured, and radioactive material was scattered all over his working place. He looked for the radioactive safety officer, but he was absent from the lab, he told Brenner about the accident and Brenner was in charge of the lab, but he sent Altman away because he was too busy and suggested to him to see Sanger. When Sanger heard about what happened, he put on rubber gloves, and with a sponge and some detergent started collecting the stuff off the floor. Altman considered this to be the ultimate demonstration of Sanger's humility and greatness. To me, it was also a demonstration of Sanger's pedagogy by showing to Altman how to handle such an

---

[134] Altman, S., "MRC LMB — April 2003." Talk by Sidney Altman to the MRC LMB meeting commemorating the 50th anniversary of the double helix discovery, Cambridge, April 26, 2003. I am grateful to Sidney Altman for a printed copy of his talk.

Frederick Sanger at the time of his first Nobel Prize, 1958 (courtesy of Frederick Sanger, Cambridge, UK); in Cambridge in 1997 (photo by the author); and with Walter Gilbert in Stockholm in 2001 (photo by the author). Walter Gilbert and Frederick Sanger shared half of the Nobel Prize in Chemistry for 1980 "for their contributions concerning the determination of base sequences in nucleic acids."

accident, which is, taking action rather than looking for others to take action. In my correspondence with Sanger, when the topic of his attitude came up, he signed his next letter to me as "Fred Modesty Sanger."[135]

> *I have asked Crick about Brenner's missing Nobel Prize. He wrote me that Brenner could've won the Nobel Prize for several things, but it would've been very difficult to single out the one for which he should have won it.*
>
> [Watson:] Yes, he could've, for the co-linearity of the gene. Francis and I, the only time we nominated together was Sydney, [Seymour] Benzer and [Charles] Yanofsky. This was for work in genetics between 1953 and 1962, the co-linearity of the gene, the polypeptide chain. Conceptually they could've given it to him, but they never did. That was the cleanest thing. Sydney's other things you couldn't pick out, except the *C. elegans*, which was very important. I tried to say that the LMB should nominate Sydney and [John] Sulston and [Robert] Waterston for the Nobel Prize for the worm.

---

[135] Sanger's letter of January 26, 2002, to the author.

Seymour Benzer at Caltech in 2004; Sydney Brenner in Cambridge in 2003; and Charles Yanofsky at Stanford University in 2006 (photos by the author).

But Max Perutz hates Sydney, because Sydney was not nice to other people.

He'll charm you. If you read his autobiography...

I read a very long interview with him,[136] *which Richard Henderson*[137] *gave me. It's very impressive.*

In the beginning, but by 1970 it doesn't have anything else.

When I later asked Brenner about Perutz, and their relationship, he told me this:[138] "It was a reasonable relationship. Max [Perutz] had two levels. There was Max Perutz the scientist and there was the Archduke Maximilian Ferdinand from old Vienna expecting a lot of things done for him. I would not say our relationship was negative. When I took over the finances of LMB in 1977, I had to do a lot of repairing because Max had a vision of the Lab, which may not have been entirely corresponding to reality. I would not consider him a mentor; he was then parallel. The person I worked a lot with was John

---

[136] Louis Wolpert's long interview has since appeared as a book, Brenner, S., *My Life in Science. As Told to Lewis Wolpert*, eds. Friedberg, E.C., Lawrence, E. Biomed Central Ltd., London, 2001.

[137] Richard Henderson was the Director of the Laboratory of Molecular Biology in Cambridge at the time of our visit.

[138] Hargittai, *Candid Science VI*, pp. 20–39. (Sydney Brenner)

Kendrew and, of course, Francis. They were my biggest connections in the Lab."

Brenner talks about Szilard engagingly. You remember last time I asked you whether you would employ Szilard and you said, probably not because he was always two steps ahead. François Jacob's answer to the same question was that he would employ him for exactly the same reason.

Yes, when he was already successful. But the average person thinks seeing Szilard that he is not going to pull his weight on today's work. Sydney is different. Sydney is very good at no matter what it is. He just treated people very badly and people are mad at him. Therefore Sydney's autobiography can't be good because he has done so many irresponsible things that a truthful autobiography would be very embarrassing.

François Jacob said to the question whether he would give a job to Leo Szilard if he would show up at his doorstep:[139] "Probably a special job of a bumblebee in a communication system whose task would be talking with people and getting and disseminating news." When I mentioned Watson's objection to him, that is, that Szilard was always two steps ahead…, Jacob added:[140] "It would be interesting to have such a person around exactly because he was two steps ahead."

In my lecture on the Nobel Prize in Stockholm I talked about those who did not get the prize and I identified a special category, scientific visionaries. They should not have been given the Nobel Prize according to Nobel's Will, but who have nevertheless had a great impact. I singled out three people, J. Desmond Bernal, George Gamow, and Leo Szilard.

Gamow didn't get it because he died too soon. Bernal would have had to get it…

---

[139] Hargittai, *Candid Science II*, p. 91. (François Jacob)
[140] Ibid.

He could have got it for the first X-ray diffraction pattern of a protein.

Yes, and with Dorothy Hodgkin. He could have got it by 1940. Bernal probably disgraced himself by his communism. But they could've given him a Nobel Prize; Perutz, Kendrew, and Bernal. No one would have been disappointed, that's what I'm trying to say.

We have already touched upon Bernal's review of *The Double Helix*. This little known review was the more interesting because Bernal was a pioneer in launching the new field of molecular biology. Molecular biology could be assigned several starting points, and all would be relevant. The most conspicuous turning point was the discovery of the double helix and for many it was when molecular biology began. However, it was also a decisive event when Bernal performed the first ever X-ray diffraction experiment on a protein — a pepsin single crystal — in 1934. In Dorothy Hodgkin's description, "that night, Bernal, full of excitement, wandered about the streets of Cambridge, thinking of the future and how much it might be possible to know about the structure of

Left: Young J. Desmond Bernal; right: Bernal giving a speech (photos courtesy of Alan L. Mackay, London).

proteins if the photographs he had just taken could be interpreted in every detail."[141]

Perutz started his Cambridge career under Bernal and switched to W. Lawrence Bragg only after Bernal had been appointed to Birkbeck College and had left for London. Crick eventually joined Perutz, and somewhat later so did Watson. Thus, considering the Bernal to Perutz to Crick and Watson succession, it is not surprising that Crick told Bernal upon receiving the news about the 1962 Nobel Prize, "Watson and I have always thought of you as our scientific 'grandfather.'"[142]

This is how Perutz reflected on Bernal:[143] "He was a prolific source of ideas and gave them away generously. During the long, lean years when most of my colleagues thought I was wasting my time on an insoluble problem, Bernal would drop in like the advent of spring, imbuing me with enthusiasm and fresh hope. A Danish colleague once said to me 'When Bernal comes to visit me, he makes me feel that my research is really worthwhile.'" Somewhere else Perutz noted of Bernal:[144] "He had a visionary's faith in the power of X-ray diffraction to solve the structures of molecules as large and complex as enzymes and viruses at a time when the structure of ordinary sugar was still unknown, and was convinced that knowledge of the structure of living molecules would lead us to understand their function."

Bernal recognized as early as 1931 that reproduction should be one-dimensional; he also asserted that the molecule involved in it should have an axis of symmetry.[145] Both of these important features were then found in the genetic code and in the structure of DNA. Bernal could not have thought of DNA in 1931, rather, he meant proteins as the carrier of reproduction, but his suggestions otherwise proved correct a few decades later.

---

[141] Hodgkin, D. C., Riley, D. P., in *Structural Chemistry and Molecular Biology*, eds. Rich, A., Davidson, N., W. H. Freeman, San Francisco and London, 1968, pp. 15–28.

[142] Brown, A., *J. D. Bernal: The Sage of Science*. Oxford University Press, Oxford, 2005, p. 352; quoting from a letter of Crick to Bernal dated November 1, 1962.

[143] Perutz, M., *London Review of Books*, July 6, 2000, p. 35. This was in a review of the book *J. D. Bernal: A Life in Science and Politics*, eds. Swann, B., Aprahamian, F. Verso, London, 1999.

[144] Perutz, M., *New Scientist*, October 21, 1976, pp. 144–147, p. 144.

[145] Brown, *Bernal*, p. 96; quoting from Bernal's notes in 1931, for the International Congress on the History of Science.

Bernal's review of *The Double Helix* in 1968 has gone unnoticed because it appeared in an obscure periodical. The review is full of insight, and only a few of its many interesting points will be mentioned here. It is reprinted in Appendix 3. Bernal mentions Torbjörn Caspersson's contribution which was an early assertion that nucleic acids are the chief constituent of the chromosomes, which contain the genes, the carriers of inheritance and that RNA is positioned between the gene and the proteins. Caspersson's name is missing from Watson's writings although he had an indirect connection to this Swedish scientist: he was strongly advised to join his Stockholm laboratory rather than moving to Cambridge when he decided to leave Denmark in 1951.[146]

Bernal mentions the lack of communication between those who worked on similar problems even in the same laboratory, which could have been avoided. Bernal was isolated from much of what was going on at the time because of his engagement in communist politicking. But there was isolation between Rosalind Franklin and the Cambridge group as well notwithstanding the fact that her data were revealed to them without her knowledge. Bernal makes an important point here because scientists, even the greatest of them, cannot function well for long in isolation. Linus Pauling's failure in the story of the discovery of the double helix was also at least partly due to his isolation. It was partly self-inflicted, but it was also during this period that he was denied his passport and could not attend a protein meeting in England because of his leftist politics. Watson and Crick were not isolated at least among themselves and the crowded conditions in the Cavendish Laboratory exposed them to visiting scientists. Watson made much use of talking with people rather than studying the literature. Thus, for example, he learned about the keto-enol forms of the bases from Jerry Donohue and about the base equivalence from Erwin Chargaff.

We have looked through the Archives through 1950 and there was only one nomination for Bernal. There was also one for Szilard.

---

[146] Judson, *Eighth Day*, p. 97.

We did not see any for Gamow. Thus the scientific community did not expect the Nobel Prize for them.

But after the discovery of the remnant heat in the Universe, Robert Wilson and [Arno] Penzias got it. If Gamow had been alive, he would've shared that. He was way ahead.

This is the problem, to be ahead of your time.

When they didn't give it to Fred Hoyle, it was a disgrace.

Watson later said of Gamow that "Geo was much a wiser individual than I first judged him."[147] But he did not recognize his contribution to molecular biology as too great, notwithstanding the book in which he copiously quoted Gamow in the period of the search for the genetic code.[148] In my letter of August 8, 2003, to Crick, I asked him about Gamow's contribution: "I have had the impression that the molecular biologists did not quite appreciate his ideas for the genetic code. On the other hand, the backs of the photographs I had received from the University of Colorado on Igor Gamow's behalf indicated Gamow as 'the originator of the triplet code.' In my conversation with Arno Penzias, when we considered Gamow's place in science history, he told me that Gamow was a better scientist than Galileo."[149]

Crick's response of August 29, 2003, included his comments about Gamow:

"I liked him very much, especially as he was very kind to two such junior scientists as Jim and me. He did not originate THE Triplet Code — his triplet code was completely wrong. I am not sure that he was the first to introduce the idea of triplets. [Cyril] Hinshelwood had a silly argument for pairs, but it's possible Dounce had earlier suggested triplets.

---

[147] Watson, *Genes, Girls and Gamow*, p. xxv.

[148] Watson, *Genes, Girls and Gamow*.

[149] Hargittai, *Candid Science IV*, pp. 272–285. (Arno Penzias; Penzias was co-discoverer of the residual heat in the Universe that gave proof for Gamow's Big Bang theory of the origin of the Universe. Penzias and his co-discoverer, Robert Wilson received the Nobel Prize in 1978.)

Left: George Gamow (courtesy of Igor Gamow, Boulder, Colorado); right: Francis Crick, Alex Rich, George Gamow, Jim Watson, and Melvin Calvin at CSHL (courtesy of James D. Watson, Cold Spring Harbor, New York).

I would rank Galileo far above Gamow, because he was the first real scientist (with the possible exception of one or two Greeks, such as Archimedes). That is, he both did experiments as well as mathematics (or quantitative thinking) as opposed to thinking in words, as for example Aristotle did. Aristotle made many perceptive observations (not all completely correct, however) but he never did an *experiment* to test his ideas. When Newton said he was 'standing on the shoulders of giants' one of the people he had in mind was Galileo. Galileo's trouble with the Catholic Church has been exaggerated, and was mainly due to the Inquisition, a quite inexcusable institution.

I will not otherwise comment on Gamow's place in modern physics. I certainly think he was original."

Gamow's role has intrigued me, especially his wanderings among the great modern areas of science of the 20th century. I talked about him with Sydney Brenner, according to whom:[150]

"Gamow defined the problem although Jim and Francis had thought about it and I had thought that it was a one-dimensional

---

[150] Hargittai, *Candid Science VI*, pp. 20–39. (Sydney Brenner)

sequence that could be translated into a three-dimensional structure."

"Gamow wrote his first letter to Jim about what was to become the genetic code soon after the announcement of the double helix, in June 1953."

"That's right. However, I went to see Francis in April 1953, before their paper appeared, we were already talking about what came out in their second paper, which appeared in May. We talked about some way to translate the DNA information into the amino acid sequence. What Gamow did was to propose a form of the code. He introduced a kind of terminology with which one could begin to discuss. In fact, everything that he did was wrong."

"What, specifically, was wrong?"

"He defined the problem; he took the view that the amino acids were assembled directly on the DNA in what he called the diamond-shaped cavities. That was his physical model, but the big mistake about this was that he did not realize that DNA has a polarity; it has a chemical polarity that reads in one direction. There is only one message because the second strand will be derived from the first by the rules of complementation. Gamow thought that you could read DNA equivalently in either direction. That was one of his degeneracies."

"Alan Mackay, who was one of J. D. Bernal's disciples, has told me that Bernal recognized early on, in the 1930s, that the genetic code could not be three-dimensional, it could not even be two-dimensional, it had to be one-dimensional, and it had to have two-fold symmetry."

"About DNA?"

"It was not yet known whether it was DNA or proteins, he was referring to the genetic material. In fact, he thought it was protein."

"The question is why did he say it should have two-fold symmetry?"

"For complementarity, I suppose."

"Pauling and Delbrück also wrote about complementarity in 1940, but then forgot about it. The ideas about complementarity in replication were around. Pauling had forgotten about it and Watson probably never read it and I had not either. Neither had Watson read von Neumann's paper on self-reproducing machines, which I did. I read that before I came to England. Anyway, that doesn't matter anymore."

"It's interesting."

"It's very interesting."

The 2004 Nobel laureate physicist David Gross found Gamow's contribution to molecular biology very important. He said that "Jim Watson distorts history a lot, but Francis Crick gave Gamow a lot of credit. Gamow didn't get it right, but he played an enormous role as a catalyst."[151]

Returning to the Watson conversation,

With so many mistakes, the Nobel Prize still has a tremendous prestige.

In biology, they made the right choices this time [referring to the 2001 prizes],[152] and this happens if you take an important field and give it to the right people. In cases like the nitric oxide, there will always be controversy over that one.

I am not an expert but the people I respect would have definitely included Moncada.

Yes. Furchgott would've got it.

---

[151] Ibid., pp. 838–855. (David Gross)
[152] The 2001 Nobel Prize in Physiology or Medicine was awarded to Leland H. Hartwell, R. Timothy Hunt, and Paul M. Nurse "for their discoveries of key regulators of the cell cycle."

Of course, he was the pioneer. He told me that he would have included Moncada.[153]

Probably Murad would've not got it. That was the Texas Mafia. The Nobel sometimes follows the Lasker.

Why is it so influential?

Because the Swedes don't have enough self-confidence. Swedish science isn't that good. It's a very tough thing.

I was surprised when the Swedish presenter mentioned the Lasker for one of the new Nobel laureates, as something very important.

I know.

She shouldn't have mentioned it.

She shouldn't have. There are no major Swedes now.

In a way, having no very great scientists might help remain objective in their judgment because they personally did not have anything immediate at stake.

Look, they do a pretty good job. My impression right now is, just being there, that Sweden is like the University of Wisconsin.

Which is quite strong.

Yes, it's one of America's better universities. It's good, and they have a lot of good people, but it's a small country.

It's still better than trying to set up a world institution...

No, no, no, no. But why did they make the mistake on Lise Meitner? Who was responsible for that? Was it anti-Semitism?

I don't think it was anti-Semitism per se, but the general atmosphere was there. One of the main characters in the Swedish Nobel institution, Hans von Euler-Chelpin ended his letters with Heil Hitler! In the 1940s.

They chose Hahn, but not Meitner; very strange.

---

[153] Hargittai, *Candid Science II*, pp. 578–593. (Robert F. Furchgott)

Lise Meitner Medal of the Royal Swedish Academy of Sciences (photo by the author).

Lise Meitner's absence from the roster of Nobel laureates is, of course, conspicuous. Otto Hahn received the Nobel Prize in Chemistry in 1945, a mere few months after the atomic bombs had been dropped over Hiroshima and Nagasaki. It was a quick recognition of the discovery of nuclear fission, which happened experimentally at the end of 1938 and was theoretically understood in 1939. By the time of Hahn and Strassmann's pivotal experiment, Lise Meitner had to flee Germany as the annexation of Austria by Germany had made her a German citizen and thus subject to anti-Jewish legislation.

In the first war years, they did not announce Nobel Prizes in the sciences; the prizes for 1943 were announced in 1944. There was a strong opinion in Sweden that they should delay giving the award for the discovery of nuclear fission for a while even after the war had ended. The progress in nuclear physics and nuclear chemistry could not be assessed realistically until more information became available from the Manhattan Project. Nevertheless, when the deliberations of the Royal Swedish Academy of Sciences were over in 1945, it was announced that Otto Hahn alone would receive the chemistry prize for the discovery of nuclear fission (for the year 1944). Although the deliberations were behind closed doors and there are no minute-taking allowed, much has become known about this case from other sources. One of the most autocratic members of the Science Academy strongly argued for giving out the prize for nuclear fission, for it being a chemistry award, and that it should go to Otto Hahn alone.[154] This suggestion won the vote by a narrow margin. Few would doubt today that Lise Meitner should have been awarded the Nobel Prize and the omission was compounded by Hahn's subsequent behavior of belittling Meitner's contribution. He did this ostensibly as part of his efforts to strengthen German science, which was much in need of his success story following Nazism and the war. The Nobel Prize institution, however, suffered a long-time damage from the injustice.

Although many deserving scientists have been missing from the Nobel roster, Lise Meitner's case has been haunting the Swedish prize-givers. In 1999, they asked a physicist member of the Royal Swedish Academy of Sciences, Ingmar Bergström, to review the case, which he did and reported his findings in a big lecture to the Academy. His main conclusion was that Lise Meitner's omission was not a case of anti-feminist and anti-Jewish discrimination, but merely one of the many who might have deservedly received the Nobel Prize, but did not. To demonstrate the absence of anti-feminist and racist sentiments of the Nobel Prize institution, Bergström stressed that Marie Curie received two Nobel Prizes and that about a quarter of Nobel laureates have been Jewish. Besides, Bergström added, "Meitner was baptized as a Lutheran and was not at all a Jewish believer."[155] Of course, the overall record of the Nobel Prize with regard to female awardees is very poor, a situation whose correction should be started at the level of universities and science academies with similarly poor records in this respect. From the report it seems as if the Nobel Prize institution would be keeping a tab of the ethnic/religious roots of the laureates, which would be very surprising. In this respect, of course, Meitner's conversion is irrelevant in the first

---

[154] I am grateful to Anders Bárány, former long-time secretary of the Nobel Committee in Physics, for this information. Unpublished records of a conversation in 2000. The story is documented in a letter from Oskar Klein, who was a participant of the deliberations of the Science Academy, to Niels Bohr, and is kept in the Niels Bohr Archives in Copenhagen.

[155] Bergström, I., letter of October 9, 2000 to the author.

place. Bergström's report has appeared in Swedish in the annual report of the Royal Swedish Academy of Sciences although in a revised variant, which diminished the controversial character of the story.[156] Lise Meitner achieved an early distinction in Sweden when she was elected a member of the Royal Swedish Academy of Sciences, the second female member of this institution. In 1999, the same year when Bergström gave his lecture on her, the Academy issued a medal honoring her. The Academy issues a medal annually, but this was the first time that a female scientist was so distinguished.

After the Lise Meitner story, a lighter topic came up in the Second Conversation; it was related to Odile Crick.

Who made the drawing of DNA for your April 1953 paper?
Odile Crick. This is mentioned in *The Double Helix.*

It is very proportional. Did you give it a lot thought?
No.

Watson stressed in his book *Genes, Girls and Gamow* that Odile Crick used "her artistic talents to draw the intertwined, base-paired, polynucleotide chains."[157] Magdi and I liked this story because it had a romantic aspect to it; Odile Crick's artistic talent making the double helix discovery into a family effort. On the occasion of our visit, we asked the Cricks about this drawing. Francis told us that the sketch was given to Odile and she only had to trace it. We thought that in this case he might have been a little more magnanimous. However, he told us about Odile's impact on the new book *The Quest for Consciousness* by Christof Koch.[158] Initially the title would have been *The Search for Consciousness* and they replaced Search by Quest at Odile's suggestion. Crick told us that the book was the result of their

---

[156] Bergström, I., *Årsberättelse 1999, Kungl. Vetenskapsakademiens* (in Swedish, Yearbook 1999 of the Royal Swedish Academy of Sciences), 1999, pp. 17–25. Professor Bergström has given me his English translation of both variants as well as he donated me all the materials he used in the preparation of his report.

[157] Watson, *Genes, Girls and Gamow*, p. 8.

[158] Koch, C., *The Quest for Consciousness: A Neurobiological Approach*. Roberts and Co., 2004.

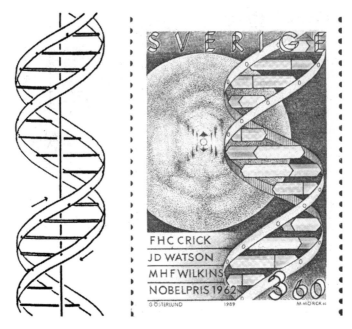

Left: Diagrammatic sketch of the double helix by Odile Crick for the original *Nature* publication; right: the double helix and Rosalind Franklin's X-ray photograph of the B form of DNA on a Swedish stamp commemorating the 1962 Nobel Prize in Physiology or Medicine.

joint work, Francis' and Christof's, but they decided that Koch should be the sole author to enhance his visibility. Francis figured only as the author of the Introduction.

Otherwise, Crick did not appear to be unduly modest. He was genuinely interested, for example, when we mentioned that other scientists refer to him more often than to any other living scientists when they are asked about their heroes. (Among the dead, Albert Einstein has been most often referred to.) He asked us what the reason might be. As an example, here I quote what Michael J. Bishop told me in answering such a question:[159] "Crick is a hero for a variety of reasons. I particularly admire how throughout his career he has maintained this remarkable ability to think about biological problems productively.

---

[159] Hargittai, *Candid Science VI*, pp. 182–199. (Michael J. Bishop)

Francis and Odile Crick with the author on February 7, 2004, in the Cricks' home, La Jolla, California (photo by M. Hargittai).

He has also been visibly removed from the kind of ambitious drive that sometimes makes scientists less attractive than they might otherwise be. I have never seen Francis driven by any ambition except inquiry and discovery."

To continue the Second Conversation with Watson:

What you should read, which is unbelievably fascinating is Brenda Maddox's life of Rosalind Franklin. It's coming out in June, Brenda sent me a copy and I've given it to Jan Witkowski. You should ask him for it. Once you start it, you'll read it to the end. I won't say anything about it.

Brenda had in her disposal Rosalind's letters she wrote to a number of people and that makes the book. You really see Rosalind through her letters.

Did she have more material than Aaron Klug had?

Yes. Aaron never had her letters. Brenda is not a scientist. The interesting thing is the person. What she was like as a person. She was what the British call upper-middle class Jewish. She was not quite at the level of the Rothschilds, but in that social world.

She looks very beautiful in some of her photographs.

In the book they point out that the family claimed to trace back their ancestors to 2000 years. It wasn't Russian Jews; it was Jews that came from Central Europe. They came to England around 1700. It was a very small community of about 20,000 people until they started coming from Russia at the turn of the century. They were very observant of Jewish laws, and so on. Anyway, Rosalind discovered freedom when she went to France and lived in a sort of left-wing communist world, and loved it. She didn't know the class distinctions.

Communism did appeal to quite a few upper-middle class people in Britain.

She didn't become a communist; it was just the intellectual community. They were communist but it was all before, when you could still say that Russia isn't bad. She found King's College very vulgar. There was a class distinction. Then she came to America and loved it. Read the book! When I went to England, my great advantage wasn't hampered by the class structure; I could do what I wanted. That's the reason that I went down to King's. Francis Crick wouldn't have intruded himself into the world of King's unless asked. It's a fascinating book. When I read her letters, I could've written the same letters, very penetrating, but she was very hard to initially make friends with. She was the enemy and she was as disliked at Bernal's place as she was at King's, except a few. She didn't have any small talk to put people at ease.

Who is the main negative character in The Double Helix?

Rosalind, because that's the way she seemed.

Not according to some. Some say that the real loser is Maurice Wilkins.

He is the tragic figure. He's tragic. When the book came out people didn't think about Rosalind, they worried about Maurice. He was a friend. We never expected to bring out the whole thing by ourselves.

He was never offered knighthood.

Knighthood is often given for working for a research council.

Not for Nobel laureates. It's an interesting collection of the Nobel laureates who did not get the knighthood. Each one of them seems to have had a reason. Josephson became entangled ...

He's nuts.

Hewish was accused of ...

Jocelyn Bell.

Maurice Wilkins because of Rosalind Franklin.

Just read the book. You'll see it. The villain was Franklin in the way she treated Maurice and the disgusting person was [John] Randall. He gave Rosalind the problem, took it away from Maurice; he was awful.

You see those pages that look like Christmas cards. There's a company in India, which is going to use bacteriophages to kill bacteria. It's back! It's back! Bacteriophages mutate so they overcome bacterial resistance much faster than the antibiotics. It may be that people went to purify themselves in the Ganges because the Ganges was filled with phage. If you ask how the Indians could be so stupid to go into that water, the answer is phage. There was a very intelligent Indian who came to my office; we're going to have a meeting on phages here, and other people come too. There's probably a perfect way to kill bacteria in a giant chicken farm and pig farms. It may come back and that was *Arrowsmith.*

Concerning Brenda Maddox's book, I received the manuscript from Watson in Cold Spring Harbor at that time in the winter of 2002, read it carefully and gave my comments to Jim, who forwarded them to Brenda Maddox.[160]

---

[160] This is how my name appeared (with my first name slightly misspelled) among the acknowledgements; Maddox, B., *Rosalind Franklin: The Dark Lady of DNA.* HarperCollins, London, 2002.

Watson's comment on bacteriophage in the Ganges was the end of the second recording.

In early spring of 2001, I asked Francis Crick about Sydney Brenner's then missing Nobel Prize. I knew that Brenner and Crick used to work together in Cambridge and that they had a sizzling intellectual interaction for years. However, I felt that from the point of view of the Nobel Prize, theirs was an asymmetric relationship. Whereas Crick had already had his Nobel Prize, the assignment of any major research achievement to Brenner might have been hindered by his close relationship with Crick. Here is what Crick wrote me on April 13, 2001:

"Although Sydney Brenner and I shared an office for 20 years, for most of that time I worked in the office (not always the same office) whereas Sydney worked mainly in the lab. However we did talk together for an hour or more on most days.

The adaptor hypothesis was my idea, but Sydney coined the name for it. Sydney had the idea that acridine mutants were probably the addition or subtraction of bases. I did all the initial work on the phase-shift mutants, but Sydney designed the special genetic cross to show that +++ mutants were like wild-type. I worked out that shifts to the left were different from shifts to the right. Sydney did almost all the work to establish the stop-chain codons. Sydney realized that the Volkin-Astrachan DNA was really messenger RNA, though I immediately saw it too. Sydney, with Meselson and Jacob, established the existence of mRNA experimentally. Sydney (and another group) established experimentally the co-linearity of gene and protein. My recollection is that all this is fairly accurately described in Horace Judson's book *The Eighth Day of Creation*.

All the initial work on the nematode was conceived and carried out by Sydney, and he organized the study of its cell lineage and its detailed neuroanatomy.

In my opinion Sydney ought to have the Nobel Prize but although he has done a vast amount of important work it is difficult to select just one particular discovery that would attract a Nobel Prize.

However Sydney's work is widely recognized by everyone. In fact he has received every other important award other than the Nobel — many more than I have!"

This letter was only 18 months before Brenner's long-awaited Nobel Prize was announced, which he shared with H. Robert Horvitz and John E. Sulston "for their discoveries concerning genetic regulation of organ development and programmed cell death."

Brenner described his interaction with Crick at LMB when in 2003 I asked him the following question:[161]

"You shared an office with Francis Crick for 20 years. Wasn't there a psychological barrier for you to overcome in discussing high-level theory with Crick and then going back to the lab and doing meticulous experiments for years?"

"No, it was not quite like that. It was just an exaggerated description of how we worked. It is also an exaggerated view that one works out the theory and formulates an experiment and then goes to the lab to carry out the experiments. This is the old Popperian idea. The other extreme is what people are now doing working on experiments, collecting natural facts. It doesn't work like that. You start by collecting facts, do a little bit of theory, and then you come back and collect more and so on. It's a play between two things, how history of science views this. What I think the most important thing to do in the lab is to interpret the experiment correctly that does *not* work. Many people today have no capacity, many of the younger people, to analyze why an experiment did not work."

---

[161] Ibid.

Brenner was not bashful about his Nobel Prize. The story is told that he was met by reporters with the news of his Nobel Prize at a European airport and his immediate response was, "It was about time." In my conversation with him at King's College in Cambridge in 2003, a few months after his Nobel Prize, I asked him whether had he received the Nobel Prize, say, 30 years ago, as he might have, would he have had the same research career during the past 30 years as he has had? According to Brenner, nothing would have been different, and he added:[162] "In fact, to me, this is my second Nobel Prize. I just failed to get the first one."

Nonetheless, when Brenner did get the Nobel Prize and was asked to deliver the traditional two-minute speech at the Nobel Banquet, he was gracious. He used the opportunity to give the following recipe for winning the Nobel Prize:[163]

"First you must choose the right place to work with generous sponsors to support you. Cambridge and the Medical Research Council will do. Then you need to discover the right animal to work on — a worm such as *C. elegans* for example. Next, choose excellent colleagues who are willing to join you in the hard work you will need to do. How about John Sulston and Robert Horvitz [his co-winners] for a starter. You must also make sure that they can find other colleagues and students. Everybody will have to work hard. Finally, and most important of all, you must select a Nobel Committee which is enlightened and appreciative and has an excellent chairman with unquestioned discernment."

Brenner counts that he could have received a Nobel Prize for all the molecular biology; for messenger RNA, the Code, for which he received his first Lasker Award. He thinks that he should have shared his first Nobel Prize with Seymour Benzer for their work in molecular

---

[162] Hargittai, *Candid Science VI*, pp. 20–39. (Sydney Brenner)
[163] Brenner, S., Banquet Speech, December 10, 2002, Stockholm (as amended by Dr. Brenner in his message to me on January 24, 2003).

genetics. Benzer did not get his Nobel Prize though he was awarded the Crafoord Prize. It has half of the monetary value of the Nobel Prize and it covers fields that the Nobel Prize does not. It is also presented by the Swedish king, but the occasion lacks the festive circumstances of the Nobel Prize. It is a big award, but little known, and the book that immortalized Benzer, *Time, Love, Memory*[164] did not even mention it in its Index. On the other hand, once it is given to someone, it may as well be considered a message that there will be no Nobel Prize.

---

[164] Weiner, J., *Time, Love, Memory: A Great Biologist and His Quest for the Origins of Behavior.* Alfred A. Knopf, New York, 1999.

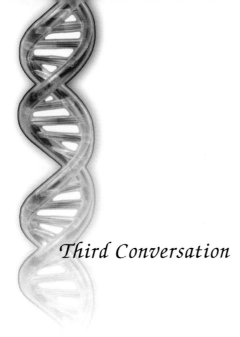

Third Conversation

The world does not belong
to the meek.

James D. Watson

As Magdi was then already working on her long-term project of
women scientists she had watched two video recordings by Mila
Pollock, the head librarian of CSHL with former women students of
Jim Watson. She thought that a conversation with Jim would be illuminating. They designated half an hour for the recording, which took
place on March 14, 2002, again in Jim's office.

> *Both of your famous former students, Nancy Hopkins and Joan
> Steitz talked about you as a promoter of women in science. Fifty
> years ago you had your encounters with Rosalind Franklin. Had
> you by then formed an opinion of women in science?*

I had my encounters with Rosalind basically through Maurice
Wilkins. There is this biography of Rosalind coming out by
Brenda Maddox, which explains a lot I didn't know. Francis
and I were friends of Wilkins and Rosalind didn't really want
to collaborate with Wilkins and certainly didn't want to collaborate with us. She didn't really welcome our interest in
DNA at the time.

Was she really as frightening as Wilkins described her?

I think it's fair to say that most people found her at King's, who didn't work with her, rather cold. This was true afterwards when she was at Birkbeck College, in Bernal's lab. Those who were not part of her group found her cold.

Was she also aggressive?

She was aggressive when she was approached. I think you'll find it worthy to read Brenda's book.

I've read it already, the preprint that is being circulated here.

Yah.

I've talked to about 50 women scientists, all of them top level, and many of them said that at one time or another they had been accused of being aggressive. The word was used with a negative connotation while, at the same time, successful men are expected to be aggressive. Is there a double standard here?

No, I don't think that when men are said to be aggressive that is a complimentary thing.

No?

No.

Nancy Hopkins said that you were the first feminist she ever met. Are you a feminist?

No, not by someone's standards. I've been a strong supporter of women who intend to do science. I treated them like anyone else, maybe a little better because of sex discrimination. I don't know the definition of feminist and that's why I'm avoiding the term. But I don't know of any case when we showed prejudice in not accepting a woman applicant, nor do I know of any case when we showed any prejudice in dealing with them as far as exams or things like that are concerned. At Harvard, our department offered a faculty position to a woman. That was very early on, some time in the mid-sixties.

Left: James D. Watson and Joan Steitz of Yale University in Cambridge, England, 2003 (photo by the author); right: Nancy Hopkins as student with James D. Watson (courtesy of Nancy Hopkins, Cambridge, Massachusetts).

Do you think that women are taken as seriously as men in science?

I don't really want to answer that because I'm not in a department, for instance, I'm not in a Harvard department, so I can't say how departments run. At Cold Spring Harbor, most of our women staff, with one exception, is very recent, and I don't go to faculty meetings. I haven't been to faculty meetings here for ten years. So I can't say anything about it only to the extent of my knowledge. You'd have to ask others. I think the situation is inherently made awkward by sexual harassment charges. In the case of men you can assume that you cannot promote someone and that's the decision. In case of a woman even though you think she is incompetent, you feel pretty sure that in many cases lawyers will appear and accuse you of harassment. That's what makes you reluctant to hire women because you have the feeling that you can't essentially fire them.

Does it mean that positive discrimination would not be good for women?

It makes me reluctant because I can be fairly honest and tough to a man, but if I feel that way to a woman, it is bad, though not as bad as with blacks, but it is halfway there.

How about black women? They are not in science.

Not yet. Since I don't do any hiring, I really have no say in this. It would be absurd, for instance, to say that half of the faculty should be women. In that case, since the pool would be smaller, those women, some of them, would be of lower quality. Life is never fair. Even if you're dealing with a society of men, there's a lot of unfairness. I don't think there was ever any prejudice against any woman I know. There were a couple of woman students that were not really very good.

You had some successful women students.

The two really successful women in science from my lab were Nancy Hopkins and Joan Steitz. Women scientists were not much part of the scientific scene until the 1970s and by that time I was not an active scientist. I was running this place.

Did you know Barbara McClintock well?

No. She was a great scientist but I never was interested in corn and I never really followed her work.

She complained to Nancy Hopkins that she would never understand the man/woman in science problem. She thought it had to be biological because she could never understand why women were not taken as seriously in science as men. She was bitter about it.

She existed at a time when women were not in faculty positions. She grew up as a poor orphan. It was lucky she got the position here.

Nancy Hopkins also said that she had made a study of women in science, which had made it to The New York Times, *and it was the first topic on which the two of you disagreed. When you read it you told her that all it was not true because — you said — if*

Barbara McClintock at CSHL (photo by and courtesy of Karl Maramorosch, Scarsdale, New York).

you want to be a great scientist, you have to be shit and women are not.

You remember that? Yah, I've said that. It's in the sense that you have to be aggressive. The same is if you look at the politics in a hospital, a group of surgeons, who's in charge, and so on. The world does not belong to the meek.

Watson did not underestimate his women students and associates in spite of such "anti-feminist" statements as when he talks about a party "at which girls, as opposed to intellectuals and their wives, would be present."[165]

His ignorance of McClintock's research may have stemmed for his lack of interest in corn; it may have also been that he did not recognize the importance of her experiments. The most Watson might have come in contact with McClintock's work was when he used to trample through her cornfield, chasing softballs.[166]

---

[165] Watson, *Genes, Girls and Gamow*, p. 51.
[166] Weiner, *Time, Love, Memory*, p. 151.

Steitz and Hopkins are Watson's conspicuously successful former students. Joan Steitz is a Professor of Molecular Biophysics and Biochemistry at Yale University. She assisted researchers in the laboratory and eventually became Watson's graduate student at Harvard University, but did not expect a university career. Her postdoctoral work at MRC Laboratory of Molecular Biology in Cambridge, England, and the changing scene in American academia with respect to women in the 1970s, led her to a premium appointment where she is emulating much of what she had learned and experienced in Watson's lab.[167] Nancy Hopkins has a prestigious professorship at the Department of Biology at MIT. Her current research interest is in the early development in zebrafish and in the role of the genes of the young zebrafish in the longevity and cancer predisposition of the adult fish species. Both were Watson's students at Harvard University and he talked about them with understandable pride.

However, Watson made no mention of Louise Chow in this conversation despite its focus on women scientists, and did not mention her either on previous occasions, when the discussion involved women scientists who had been overlooked in the Nobel Prizes. Although Chow had been very successful at CSHL, to take account of her achievements may raise painful questions for Watson. They go back to the 1993 Nobel Prize in Physiology or Medicine and before. This Nobel Prize was given to Richard Roberts and Philip Sharp. In the presentation speech at the award ceremony, the Swedish academician, Bertil Daneholt of the Karolinska Institute spoke exclusively of Roberts and Sharp as the two independent discoverers of split genes in 1977.[168] There were not only no other names mentioned, but there was not even a hint that others might have been involved in the discovery. This is a practice often followed by the Nobel institutions: they overemphasize the contributions of the awardees and underplay those of

---

[167] Steitz, J., "Flowers and Phage." In *Inspiring Science*, pp. 159–165.
[168] *Nobel Lectures: Physiology or Medicine 1991–1995.* World Scientific, Singapore, 1997, pp. 117–118.

Left: Richard Roberts in Cambridge, 2003; right: Philip Sharp in Stockholm, 2001 (photos by the author). The two shared the Nobel Prize in Physiology or Medicine for 1993 "for their discoveries of split genes."

others. This comes out as if adding credibility and justification to the decisions about the prizes, and appears as an attempt of rewriting science history.

There is no doubt though that the discovery of split genes deserved the highest award. Before this discovery, the general view was that the genetic information in DNA is continuous. After the discovery, we now know that only segments of DNA contain the genetic information. These segments are called exons and they are interrupted by the so-called introns that carry no genetic information. The origin and function (if any) of the introns remain unknown to date. The gene is fully transcribed into RNA, but in the next step, messenger RNA contains only the exons stitched together after the introns had been excised.

Roberts spent the portion of his career at CSHL during which split genes were discovered. Even Sharp spent a postdoctoral stint there, but his major contribution was made at MIT where he had moved after CSHL. Roberts, however, had moved to the private company in Massachusetts, where he still is, only after his work related to the discovery had been completed at CSHL. Roberts was the principal researcher in this work at CSHL, without having been its overall director. Others participated in the work, amounting to about a dozen workers in more or less senior positions. There was a crucial experiment at CSHL using electron microscopy by Louise Chow and Thomas R. Broker, a wife and husband team. Roberts suggested the initial experiment, but Chow and Broker refined and carried it out. Although Roberts's hypothesis was not proved, the electron microscope experiments came up with the decisive evidence about split DNA from which it became unambiguously established that genetic information in DNA is not continuous.

Roberts had compiled a 15-page document about the history of the discovery of split genes in which he allegedly downplayed the role of the electron microscope work.[169] Watson used this document to lobby for the Nobel Prize for the discovery on Roberts's behalf. However, Watson later complained that "Louise did it, and it's terrible that she didn't win it."[170] The problem might have been that at MIT, in addition to Sharp, Susan Berget — also for electron microscope work and also a woman scientist — could have been included, making the potential awardees a foursome, which is impossible according to the bylaws of the Nobel Prize. They limit the number of awardees to three for a given year in the same category. It is not clear whether Watson has developed his regrets about Chow's absence from the Nobel roster in hindsight, or whether he considered her contribution deserving the award from the start. The latter would not be consistent with his using Roberts's compilation for his lobbying efforts. In his turn, Roberts did not go out of his way to give too much credit to Chow's and others in his Nobel lecture. It sounds almost condescending when he writes that "Louise Chow and Tom

---

[169] The question of credit for the CSHL side of the discovery has been much discussed and was used as lead topic in an analysis of assigning credit for scientific discoveries titled soon after the Nobel Prize: Cohen, J., "The Culture of Credit." *Science* 1995, 268, June 23, pp. 1706–1711.
[170] Ibid., p. 1708.

Broker, two talented electron microscopists, agreed to collaborate with us on the crucial experiment."[171]

Watson's emphasis on the importance of being aggressive, alluded to above, is consistent with his views[172] on the need of an enemy in scientific research because it motivates; enhances competitiveness; and generates anger, which stimulates. The beneficial effects of having enemies have been noted by others as well, and it is not pure irony. The Nobel laureate chemist George A. Olah, referring to another Nobelist, George von Bekesy, says, "What all scientists need is to have a few good enemies. When you do your work and write it up and you send it to your friends, asking for their comments, they are generally busy people and can afford only a limited amount of time and effort to do this. But if you have a dedicated enemy, he will spend unlimited time, effort, and resources to try to prove that you are wrong."[173]

Watson's views on what facilitates science got reflection in the atmosphere he created at Cold Spring Harbor Laboratory at least at the time of his active involvement in directing it. Ellen Daniell is writing about her experience as a postdoctoral scientist at CSHL in the mid-1970s, that is, at the time when Watson was fully in charge. Daniell arrived at CSHL from the University of California at San Diego (UCSD) in La Jolla, where she had obtained her PhD in bacterial virus work. She characterized her years at UCSD as "a wonderful experience."[174] She then spent two years as a postdoctoral fellow at CSHL.

This is how she described her experience at CSHL:[175] "Cold Spring Harbor was a big change from university life. It was a small establishment, and everyone lived on the grounds or very nearby. People talked about science all the time or gossiped about the private life of others. My adviser, who worked long hours, expected to see everyone in his group back in the lab in the evenings.

---

[171] *Nobel Lectures: Physiology or Medicine 1991–1995.* World Scientific, Singapore, 1997, p. 124.
[172] McElheny, *Watson and DNA*, p. 102.
[173] Hargittai, *Candid Science*, p. 275. (George A. Olah)
[174] Daniell, E., *Every Other Thursday: Stories and Strategies from Successful Women Scientists.* Yale University Press, New Haven and London, 2006, p. 39.
[175] Ibid., pp. 41–42.

One colleague was ridiculed because he babysat for his young daughter one or two nights a week so his wife could pursue her career in the performing arts. The directors fostered competition between 'The Lab' and groups at other places doing similar work. I had seen competition before, but this felt nastier. When someone returned from a conference and reported that investigators at the National Institutes of Health were on the wrong track about some phenomenon our scientists had figured out, team members seemed to take more pleasure from the competitors' error than from their own achievement. Competition among people within the institution was also nurtured. It was not unusual for two postdoctoral fellows to discover that they were working on the same scientific problem, one or both having been assured that it was solely his or her project to develop and pursue. The heads of different units vied for recognition, derided one another, and some forbade their research fellows to talk to those in the other labs. In this environment, I developed a new set of feelings about science. I was beset by doubts about my ability and was no longer certain that my devotion to science was sufficient to achieve success I craved and had anticipated."

This was perhaps the atmosphere at CSHL about which Horace Judson remarked that under Watson, it became "a powerhouse of biology, a place where the young Jim Watson could hardly have flourished."[176]

Speaking about enemies, Edward O. Wilson, the Harvard biologist and Pulitzer-Prize-winning author writes in his memoir, in a chapter titled "The Molecular Wars" that he has been "blessed with brilliant enemies. They made me suffer (after all, they were enemies), but I owe them a great debt, because they redoubled my energies and drove me in new directions." At Harvard, he encountered Watson who "served as one such adverse hero for me."[177] Wilson's further characterization of Watson is quoted in Appendix 2.

[176] Judson, H. F., *The Eighth Day of Creation: Makers of the Revolution in Biology*, Expanded Edition. Cold Spring Harbor Laboratory Press, Cold Spring Harbor, New York, 1996, p. 613.
[177] Wilson, E. O., *Naturalist*. Warner Books, Island Press, 1994, p. 218.

Apparently, Watson has thought a lot about how to succeed in science. He devoted his talk given for the 40th anniversary of the double helix discovery to this topic.[178] Fame was a driving force for him. Had he stayed with birds — as he spent the first three years of college with studying birds — no one would have heard of him, he admitted. He set up a set of rules that assured success:[179]

1. Avoid dumb people.
2. To make a huge success, a scientist has to be prepared to get into deep trouble.
3. Be sure you always have someone up your sleeve who will save you when you find yourself in deep s–.
4. Never do anything that bores you.
5. Expose your ideas to informed criticism.
6. If you can't stand to be with your real peers, get out of science.

It is of interest to compare Watson's rules for success in science with others; here we provide an abridged set of Donald Glaser's rules:[180]

- Have a fighter in the belly; a real drive to do something.
- Either know a lot or have the ability to learn a lot about a field.
- Have an ability to make random combinations quickly among the things you know.
- Have the ability to quickly reject the ridiculous combinations or work with those that promise something even though appearing ridiculous.
- Have professional competence and knowledge to work out those combinations that survive this filtration.
- Have luck; that is, know when to quit if things are not promising even if a lot of effort had been invested in them.
- Have the ability to work with others and talk to others.

---

[178] Watson, J. D., "Succeeding in Science: Some Rules of Thumb." *Science* 1993, 261, September 24, pp. 1812–1813.
[179] Ibid.
[180] Hargittai, *Candid Science VI*, pp. 518–553. (Donald Glaser)

The most conspicuous difference between the two sets is Watson's emphasis on risk-taking and reliance on possible rescue by others.

Watson's keys to success are to be found, however, in a broader domain of traits than his six points. Knowing that he had the ability to distinguish between the important and non-important, it is remarkable that he always found time for relaxation. He appeared to be a person economizing with his time, but not when he was doing something that he judged truly needed. Thus he was very patient when he was cutting out his paper models of the bases as he was on the verge of the discovery of base-pairing. He paid meticulous attention to the minutest details when writing his textbooks. He devoted a lot of time to the back-and-forth exchanges with his colleagues and friends as he was preparing the publication of his book *The Double Helix*. He has paid again the most careful attention to the minutest details in the planning of new constructions and renovating old buildings at Cold Spring Harbor Laboratory.

It is also important what he did not do. There are scientists who once they find a fertile area of research, they exploit it to the fullest; once they establish a new methodology, they apply it to whatever it may be applicable. Others may feel in retrospect that they had moved on too quickly after they had made a discovery. For some it is a real question, "how you would stay in the groove long enough and get out before you are in the rut."[181] For Watson, it was never a problem to determine when his work became repetitious without, however, having utilized the potentials of an area. After the discovery of the double helix and after having made it sure that everybody saw that Watson and Crick understood its biological implications, he moved on. His experience with the study of the structure of RNA and with the quest for the messenger RNA further must have strengthened his determination that instead of trying to top his previous feat, he should be seeking his success in other aspects of science.

Watson has been very lucky, but he has worked hard at finding his luck. He always found the right mentors; supporters; partners;

[181] Conversation with F. Sherwood Rowland; In Hargittai, *Candid Science*, pp. 448–465, pp. 450–451.

Left: Rosalind Franklin (courtesy of Aaron Klug, Cambridge, England); right: Memorial Plaque commemorating the X-ray diffraction studies on DNA at King's College in London, listing R. E. Franklin; R. G. Gosling; A. R. Stokes; M. H. F. Wilkins; and H. R. Wilson (photo by the author).

ultimately, the right wife; the right venues for remaking a research place into his own image; and most of all, the right shoulders to stand on in order to look further. In this, Watson and Crick may have "stolen" Rosalind Franklin's data (and this is debatable), but the fact is that they recognized the importance of the problem whereas she did not. Incidentally, whether it was "legal" or not to show Franklin's data to Crick and Watson, there has been much effort to demonstrate that there was nothing wrong with it.[182] It is revealing how decades later Watson referred to the episode when he went to King's College to spread the news about Pauling's erroneous triple helix model of DNA. On that occasion, in Watson's words,[183] "Maurice — bristling with anger at having been shackled now for almost two years by Rosalind's intransigence — let loose the, until then, closely guarded King's secret that DNA existed in a paracrystalline (B) form as well as a crystalline (A) form." So it was not so much altruistically sharing information for the sake of advancement of science than an angry revenge. In any case, Watson was not eager "to stir up more discussion as to whether we had improperly used the King's College data."[184]

---

[182] See, e. g., the notes by Perutz and others in *Science* 1969, June 27, pp. 1537–1538, following the publication of Watson's book *The Double Helix*.
[183] Watson, *Genes, Girls and Gamow*, p. 7.
[184] Ibid., p. 21.

Back, again, to the question of aggressiveness.

Whether there is an inherent difference in aggressive person-ality, that is, a genetic difference, between women and men, I have no idea. Women are often brought up to be less com-petitive in that sense. There is — you could say — a revival of women's athletics, to be competitive to win. Society may be changing. You have to decide you want to be first.

But we are talking about aggressiveness and not unethical behavior.

No, but the distinction of what is unethical is very arbitrary. You have to define your standards. I talked with an old friend, Arthur Kornberg, who is certainly not a feminist and what he says basically that women can never be equal in universities because they want part-time positions. He was at a medical school with 24 women and the families were opting for not working for 80 hours a week whereas men in a similar posi-tion would work 80 hours a week. I think you have to give the jobs to people who do the best science. If doing the best sci-ence is the result partly of working 80 hours a week, someone who wants to work 40 cannot compete with the one who works 80 hours. There're all sorts of schemes of how to let women back in after the children are no longer infants and things like that. I would certainly favor those sorts of things. Nancy was very strong; yet she couldn't have had children and be a competitive scientist. It's very tough.

Did she have children?

She could've, but her marriage broke apart.

Are you basically saying that the situation is hopeless for women who would like to have both a career in science and a family?

Some do it; Dorothy Hodgkin did, but she had servants. There's also daycare. You begin to do things. But the quality of really good science is pretty extraordinary obsession. It's hard to be obsessed about two things at the same time; about your family and your work. I guess I'm just saying that in a science department you could be fair and have the number of women to be half of that of men. Some will be able to arrange their lives and have the right personality for it; others will say, it's just asking too much.

If a woman came to you asking for advice, if she would like to do science and raise a family, what would be your advice?

Do work which is not experimental.

Is it easier?

I don't know; I'm just trying to say that you can't have every-thing. Women want to have children in many cases and men don't; that's their real way. You can't change the essential dif-ference between men and women. You can't make some things equal when you don't have the basis for it. Given everything in nature, it's harder for women to be in the top 1% of the field.

There are some statistics that show that as you go higher on the academic ladder, the number of women is getting smaller and smaller. For full professors the ratio is about 10%, for members of the national science academies, it is about 3% and it is about the same for the science Nobel Prizes.

I don't think there is any discrimination against women in awarding the prizes. Take, for example, Christiane Nüsslein-Volhard, she has no children, she is rather tough, and she is a superb scientist.

After the Nobel Prize, she had to see an analyst. Apparently, the tremendous tension may have led to some depression.

The mind of women operates very differently, women think differently than men. The social constraints are very different. I think you have to go out of your way to give them enough opportunities. If they don't seize some you can't promote them when they really don't want to work as hard as the men. The President of Radcliffe, a woman chemist, about 15 years ago gave a lecture and said that men should work less hard.

It wasn't fair to women to compete with men who work so hard. But I thought what should we do? That didn't make sense. She was trying to balance things.

Do you think there is a feminine way of thinking about science?
I don't know. I don't believe that.

Shirley Tilghman has just accepted the presidency of Princeton. Is it a loss for science?
No. You know that I also came here [to CSHL]. It's kind of crazy that you should do the same thing all your life. It is important that sometimes good scientists run great institutions. It's good for science because there is then someone who understands science. You could be selfish and say, yes, it's good for us.

There is still one more question about Rosalind Franklin that I would like to ask you about personally. Having read what I have found available, I still have this one question. After you saw her X-ray picture, and that was an important element in the discovery, you never told her that you had seen the picture.
I never saw her until the structure was found. There was no occasion of talking to her. There wouldn't have been any reason to talk to her. The picture — I think — was actually taken by Gosling and it was in Wilkins' possession, she was transferring it, and he showed it to me.

I understand that. What I was wondering about is that you never told her. Do you think she never learned that?
She, of course, knew that we had seen these pictures. I'm sure she knew this when she came up to Cambridge. I can't imagine that she didn't. People make a lot of it. But when it was over, she talked to us; she never said that we treated her badly. You could see in *Genes and Girls* that our relationship was very straightforward and she was certainly treated as an equal, and she had data that we shared. Brenda [Maddox] makes a lot of us not describing the crystallographic data, but I never used the dyad

[the presence of twofold symmetry] to find the structure.[185] Once we had the base-pairs we were sure of the structure. So it was not an essential thing in my mind. Until the base-pairs were found, Francis was not thinking of the problem, not in a full time sense. In the paper we mention the unpublished data of King's. There were two things. One was the X-ray photograph saying there was a defined structure and that's why I came to Cambridge. If it had been a powder diagram, I doubt that I would've made this clear change. So that was Wilkins' contribution. Then when Rosalind got the B form, and Wilkins saw it, he and Stokes thought that it was a helix. That was in September 1951 before even I arrived in Cambridge. There was a big blow up between Rosalind and Maurice in that September. She essentially rejected the helical model. Wilkins was essentially in favor of the helix while she was strongly against it. The helix was the only thing we could build anyway, but seeing that X-ray photograph even that 2% chance that it was not a helix was gone. It was a helix. Then the last time she had to write up her data, the first of February 1953, she looked at the B form and began to see it was a two-chain molecule. The solution came from model building and Linus [Pauling] should've built the double helix without ever having seen either Wilkins or Rosalind's photographs. He should've done it.

How could he make such a mistake?

I don't know. He didn't look at the data. There was the Astbury photograph that was published in 1938; it was in the Cold Spring Harbor Symposium and it looked like the B form. So I never felt that Rosalind's data was that vital because we would've built the model with a 30 or 34 angstrom repeat

---

[185] Aaron Klug considers this question in his historical perspective, "The Discovery of the Double Helix," *J. Mol. Biol.* 2004, 335, pp. 3–26. Klug points out that Watson made no explicit use of the X-ray data to find the base pairs, but Crick on seeing Franklin's contribution to the MRC report, realized that her C_2 symmetry of the **A** form implied two chains running in opposite directions. Watson had been trying two-chain models, but had them running in the same direction. I thank Aaron Klug for pointing out this to me in his letter of September 26, 2006.

September 1953 meeting on protein structures at Caltech in Pasadena, California (courtesy of Gunther S. Stent). From left, first — Maurice Wilkins (middle row); second — John Kendrew (front); fourth — Alexander Rich (middle); fifth — Beatrice Magdoff (front); sixth — Max Perutz (back); ninth — Linus Pauling (front); tenth — Jim Watson (back); thirteenth — Robert Corey (front); fifteenth — Francis Crick (middle); seventeenth — Richard Marsh (back); twentieth — William Astbury (front); twenty-fourth — W. Lawrence Bragg (front); thirty-sixth — John Randall (front); forty-fourth — George Beadle (back).

because the chain wanted to go around eight to ten times. Ours was a very different approach from what Rosalind did.

Did you ever suggest to Wilkins to be a co-author of your paper?

 I didn't. I don't remember that. Francis might have because it was Francis who talked with Maurice.

Was Wilkins' Nobel Prize justified?

Yes, because he started it.

But he did not get the result. Did he get important results after 1953?

No. Nirenberg did the important thing for the code. Getting the crystalline thing was what made the structure possible. We never got a crystalline photograph of RNA and it was not solvable. When you have a crystalline form, you know there is some structure there to be solved. Wilkins got Stokes into it, he solved the helix, and he worked out the diffraction pattern. Wilkins' fault was that he didn't tell us, "Go to hell." The real victim was never Rosalind, it was Maurice. She thought she could take someone else's problem away and never bother about it.

Did you respond to the protest that followed the publication of The Double Helix?
I didn't say anything.

Did you mind it?
I didn't care.

Did you ever have a bad conscience about the way the discovery happened?
No. Because we never expected to come up with such a beautiful answer and become so famous. We thought we would get the model and the King's data would be necessary to prove it. Then it turned out that the model was so pretty, it would've been very hard not to believe it. It was very simple. Our luck was that the others didn't get it; didn't solve a simple problem.

In your recent book, Genes, Girls and Gamow, *you talk a lot about all the girls you met but say very little about Liz.*
I don't think you write about your own marriage. I don't know anyone who has written a book about his marriage. You don't do it.

Why did you marry in secret, in California?
My father was dying. He had lung cancer. He was very sick. He was not at the wedding; there was no way he could've been at the wedding.

So you wanted to get married quickly.
You don't basically live with a student. You either marry them or you don't live with them, all right?

The Watson family on X'mas card by Lewis Miller (courtesy of James D. Watson, Cold Spring Harbor, New York).

How much time do you have for family life?

None now, for this year. For many years after I got married it wasn't as hectic as this. Having a very sick son has been a very complicating thing. It's still very complicated. A lot of our time is going to what you call non-productive ends.

I would like to ask you about your writing. Is it just a gift or the result of consciously developing this ability?

All I can say is that I've read a lot and if you read a lot it helps.

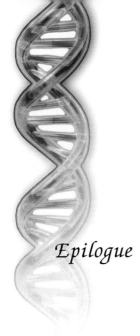

Epilogue

Worrier and Warrior, Jim has been
the guardian of DNA for the past 50 years.

Sydney Brenner

Watson has had an extraordinary career a great deal of which he had
charted, though he could not have charted the discovery of the double helix — and it having such a structure that pointed to its function.

He always had his goals before him. From the time he was halfway
through his undergraduate studies at the University of Chicago he
was striving to become a scientist, as soon as his interest shifted from
bird watching to understanding the gene. In his youth, he was more
ambitious than the other students and he was the keenest among
them.[186] He realized that there were people more talented than he
was at the university, but this did not bother him; rather, he sought
out their company, and learned from them.

Watson often modeled his behavior after others and this may
explain the different impressions different people could have formed

---

[186] Szántó, T., Interview with James D. Watson, Cold Spring Harbor, September 24 and 27,
1986. The Hungarian translation of this interview has appeared in *Magyar Tudomány*, 1987,
pp. 866–870. I am grateful to Dr. Szántó (Budapest) for providing me with the original transcripts of the interview in English.

about him at different times. This also explains why it is possible to discern in Watson the traits of other scientists; both positive and negative. He has lived a unique life, which after the discovery of the double helix was in part determined by having become so distinguished at such an early age. His time from six years old to 25 was the "normal" part of his life, but even that part compressed far too many happenings to consider it ordinary. As a young man he seems to have developed into something of a nerd. An example of his inconsiderate behavior is mentioned by Sydney Brenner; as Watson came for a visit to the Brenners, their children were having a birthday party, and Watson ate all the cake that had been prepared for the children.[187] He may have been inconsiderate in more serious things than birthday cakes, yet people kept seeking his company and returning to him, so he must have charmed them at the same time in addition of being world-famous and increasingly influential. His intelligence combined with external shyness and shabby clothing — even if it was affected — may have made him an anti-hero, who elicited sympathy and support. Brenda Maddox noted that "Watson brought out the motherly instincts in the Cambridge wives who saw that he was interested in girls, but was unsuccessful."[188] Salvador Luria tells the story of a lady condemning the director of the Cologne Institute where Watson gave a talk, appearing so weak that the lady thought that he "was obviously near death." Luria adds though that Watson "did then and still does look frail, although he is, as the saying goes, strong as an ox."[189]

I have heard personal accounts, without having seen much in print, about how people were afraid of him later in his career, and how arbitrary he could be. The Watson I knew not only in relationship to me and my wife but whom I could observe with respect to others was kind and considerate. Only whispered stories at Cold Spring Harbor referred to his imperious behavior in the past, like firing someone just because the person parked his vehicle blocking Watson's

---

[187] Brenner, *My Life in Science*, p. 42.
[188] Maddox, *Rosalind Franklin*, p. 182.
[189] Luria, S. E., *A Slot Machine, a Broken Test Tube: An Autobiography*. Harper & Row, New York, 1984, p. 130.

car. I had only one glimpse into how nasty he could be when we talked about an unpleasant review of *Genes, Girls and Gamow*. Watson tried to discredit the reviewer, but could not find much about his writing, so — as a spoiled child — he brought up personal traits of his adversary. It was an embarrassing moment.

Watson had little involvement in national politics. At one time, around 1968, he did work for the President's Scientific Advisory Committee, PSAC. When President Richard Nixon declared his "War against Cancer," Watson came out strongly and expertly about the futility of the project, pointing out that the money would be ill-spent if there were no basic understanding of the causes of the different cases and the mechanisms operating. He was proved right. Then, there was Watson's nationally prominent role in the Human Genome Project between 1989 and 2002. Apart from terrorism and the Israeli-Palestinian affairs in foreign relations, and abortion in domestic ones, politics hardly entered our conversations. This may have been my having not asked the appropriate questions, but Watson's writings do not reveal much interest in politics either. There is partisanship in his stressing of being a Democrat, which was a family tradition, but not much beyond that. In *Genes, Girls and Gamow*, he noted the involvement of some of his colleagues in politics, Linus Pauling's in particular, but he did this just as an observer. The impression is that he was more interested in political questions when they had consequences for him personally. Thus, Watson was glad that McGeorge Bundy, Dean of Harvard and later one of the architects of America's involvement in Vietnam, did not leave the university before having helped install him at the Biological Laboratories.[190] He was abhorred by official America's preventing Pauling from going to Britain for an important meeting, but felt some satisfaction that this delayed Pauling in seeing the X-ray patterns of DNA at King's College.

Pauling[191] and Watson were both extremely dedicated to what they were doing and were equally success-oriented, but differed in

[190] Maddox, *Rosalind Franklin*, p. 261.
[191] See, e.g., Hager, T., *Force of Nature: The Life of Linus Pauling*. Simon & Schuster, New York, 1995.

some important aspects, for example, in their lecturing styles. Pauling
was flamboyant; he was a true showman; there were well-designed
andantes and crescendos in his performance. Watson spoke mostly in
such a quiet voice that was hardly audible; often speaking to the
blackboard; hesitations and silences punctuated his speech; it was not
a theatrical achievement. Or was it? The result was the same in that
each kept their respective audiences spellbound.

For science, Pauling transformed chemistry by bringing quantum
mechanics into it before anybody else and he had a long-lasting
impact; long-lasting, but not eternal because his approach was eventually superseded. In describing how atoms are kept together in the
molecule, Pauling's Valence Bond (VB) Theory had an archrival in
Robert Mulliken's Molecular Orbital (MO) Theory. For a while it
seemed that the VB theory would carry the day due to its visual
attraction and Pauling's dazzling performances, but it just did that, it
carried the "day." In the long run, Mulliken's MO theory has proved
infinitely more applicable for computation and has been a success
story in chemistry. From the mid-1950s, Pauling's attention turned
to a great extent to politics, which he pursued not only enthusiasti-
cally but fruitfully as well; suffice it to mention his advocacy for the
nuclear test ban and his Nobel Peace Prize. He also dedicated himself
to scientific crusades; some proved an exaggeration, like Vitamin C,
while others, like his concept of molecular medicine, proved
prophetic, but largely after his time. Pauling was virtually forced out
of Caltech by alienated trustees who found his political activities dam-
aging for the School. Later Pauling ended up with his Linus Pauling
Institute, which, however, was of marginal importance. It struggled
for solvency, and disappeared soon after Pauling was gone. It is hard
to imagine Watson to devote himself to hopeless or lost causes, and
CSHL is conspicuously robust and expanding geographically and
financially as well as in the scope of its activities.

Watson appeared sympathetic, at least in retrospect, to Robert
Oppenheimer's plight about his 1954 security hearing, though not to
his pre-war leftist political activities.[192] Oppenheimer's mysterious

---

[192] Ibid., pp. 102–103, p. 118.

aura caught Watson's attention and he attended one of the physicist's lectures, which impressed him although he did not understand its specifics. When they crossed path on Caltech's ground, Watson felt as if he had known Oppenheimer, but this interest was one-sided.[193] McElheny[194] describes how Watson irritated his fellow graduate students at Indiana with his questions and interruptions during their seminar presentations. This reminds us of the story when Robert Oppenheimer's colleagues had to complain to Professor Max Born about Oppenheimer's similar behavior.[195] There is no record that Watson's fellow students would have filed a complaint with Luria.

When Watson had questions, he preferred to consult the professors, Muller, Sonneborn, or Luria to talking them over with his peers.[196] He "had no inhibition about going up to the great men of science to engage them in discussion, although he was but 19 years of age."[197] This is just as Leo Szilard behaved at the Berlin physics

Left: Leo Szilard (courtesy of Endre Czeizel, Budapest); right: Max Delbrück, Aaron Novick, Leo Szilard, and Jim Watson at CSHL (courtesy of James D. Watson, Cold Spring Harbor, New York).

---

[193] Ibid., p. 171.
[194] McElheny, *Watson and DNA*, p. 18.
[195] Born, M., *My Life: Recollections of a Nobel Laureate*. Charles Scribner's Sons, New York, 1975, p. 229; Pais, A., *Inward Bound*. Oxford University Press, 1988, p. 367.
[196] McElheny, *Watson and DNA*, p. 18.
[197] Olby, R., *The Path to the Double Helix: The Discovery of DNA*. Dover Publications, New York, 1994, p. 298.

colloquia in the 1920s, which he started frequenting at the age of 22 and where he engaged in conversation with such giants of modern physics as Max Planck, Max von Laue, and others.[198] Comparison with the Berlin colloquia may have entered Szilard's mind when in June 1953, Watson made his first presentation at the Cold Spring Harbor laboratory about the discovery of the double helix. In the first row were sitting such luminaries as Delbrück and Szilard.[199]

There were other similarities between Watson and Szilard. An example is found in Watson referring back to his Cambridge time, "[I] said what I thought as opposed to what good manners required."[200] For Szilard, "addiction to the truth was victorious over whatever inclination he might have had to be tactful." According to him, clarity of judgment is not a matter of intelligence; rather it "is a matter of ability to keep free from emotional involvement."[201] Both preferred truth to politeness[202] and both were outspoken without reverence to authority. Incidentally, upon their first meeting, Szilard told Watson that he should learn to speak clearly if he wanted to be understood. This was not at all unfriendly from Szilard; his scolding was an indirect appreciation because by doing so he acknowledged that Watson said things worthy of his attention. Neither minded if they were different from their peers in their early childhoods.[203] There was something else common between them; neither found it fruitful looking back; Watson stated what Szilard had as one of his mottos, "The future is what counts."[204]

For all the flamboyance many see in Watson, he strikes me as a conservative person who is careful in risk-taking despite his emphasis of risk-taking in his list of traits leading to success (see above). Again,

[198] Hargittai, *Martians of Science*, p. 43.
[199] Ibid., p. 74.
[200] Watson, *Genes, Girls and Gamow*, p. 6.
[201] *Leo Szilard: His Version of the Facts*, p. 5.
[202] For Szilard, see Hargittai, *Martians of Science*; for Watson, see, McElheny, *Watson and DNA*, p. 9 and p. 16, respectively.
[203] For Szilard, see Hargittai, *Martians of Science*; for Watson, McElheny, *Watson and DNA*, p. 56.
[204] McElheny, *Watson and DNA*, p. 182.

a comparison with Szilard is telling here. Szilard ignored hierarchy in science and, so it seems, has Watson. But Watson operated within this hierarchy, whereas Szilard was truly never interested in it. Watson's flamboyance came mostly from the security of his position whereas Szilard's from his disregarding the consequences on his position, which did not exist in the first place. Szilard ignored the expectations of society; Watson has been keenly aware of them, especially those of the non-scientific part of society. He seems to have been thriving in high society, and while he often labeled fellow scientists boring, he refrained from using such labels for societal people. Watson's interest in stocks and bonds and options has been noted;[205] they never entered Szilard's equations. Had Watson not enjoyed high society, he could have bowed out, but he did not, and has written obituaries for the benefactors of CSHL when they died. One can hardly imagine Szilard penning accolades of businessmen.

Watson believed in talking with people and he thrived on it although his partners thought him to be taciturn. Maybe he was a better listener than talker, but he recognized the importance of active scientific interactions. The prime example of this is the Watson-Crick partnership, which can be especially appreciated when the opposite situation occurs. The British James W. Black, Nobel laureate biomedical scientist, discoverer of important drugs at one point at an early stage of his research career, in 1947, found himself working at the Medical School in Singapore, then a British colony. He was building ingenious experiments studying the blood pressure-blood flow relations in the circulation of experimental animals. He had a lot of results, but found it impossible to judge their value because he had nobody to talk to about his work. He oscillated between thinking, "this is great" and "this is rubbish." He learned the importance of contacts with colleagues in order to calibrate one's intellectual activity.[206] Closer to Watson, Rosalind Franklin could not have been more isolated in Singapore than she was in London while she was working on DNA. It seems also that Erwin Chargaff did not have the right

---

[205] Weiner, *Time, Love, Memory*, p. 169.
[206] Hargittai, *Candid Science II*, pp. 524–541, pp. 528–529. (James W. Black)

intellectual environment that would have been conducive to following up his seminal observations about DNA composition, and leading
to the DNA structure. In fact, pondering about whether he might
have come up with the right model, Chargaff noted that "if Rosalind
Franklin and I could have collaborated, we might have come up with
something of the sort in one or two years."[207]

Watson seldom found himself in a situation of intellectual desert
and even then for brief periods only. When it did happen, he "madly
hypothesized"[208] about RNA chains in TMV during his postdoctoral
stint at Caltech in 1955. Watson made every attempt never to find
himself without partners, and this was characteristic of Crick as well.
Crick displayed a relevant quote on the title page of his manuscript
about the adaptor hypothesis that he circulated among members of the
RNA Tie Club. The paper suggested the existence of small adaptor
molecules that would temporarily attach to the amino acids on their
way of incorporation into a protein. Crick did not consider it to be a
publishable work, but badly needed feedback from his colleagues. The
quote referred to above said, "Is there anyone so utterly lost as he that
seeks a way where there is no way" (*Kai Ka'us ibn Iskander*).[209]

The Watson-Crick partnership is one of the most famous and
fruitful ones in the annals of science. They hit it off as soon as they
met when Watson arrived in Cambridge and joined Perutz and
Kendrew at the Cavendish Laboratory. They were different, but the
differences were superficial, and the similarities dominated. Watson
summarized them in 1986 in the following way:[210] "We both like to
think about important things. Crick does not use his brain to play
chess. We want to use our brains for understanding nature, we are
not just puzzle-solvers. We both are interested in understanding facts.
We are always seeking clues. We are rather pragmatic, if we get stuck
in one way or another one. We are not very sentimental. We're

[207] Chargaff, *Heraclitean Fire*, p. 103.
[208] Watson, *Genes, Girls and Gamow*, p. 121.
[209] Ibid., p. 165.
[210] Szántó, T., Interview with James D. Watson, Cold Spring Harbor, September 24 and 27,
1986. The Hungarian translation of this interview has appeared in *Magyar Tudomány*, 1987,
pp. 866–870.

straightforward. If there is something wrong with a theory we tell it is wrong. That is not to be nasty; of course, it is just the spirit of science. A sentimental science would not survive. Then we don't much care about what other people think of us. We are not upset if people are talking that we are strange, crazy or whatever. We were trying to do something quite different. Most people do not like to take chances. We are willing to take reasonable chances, let's say 30% or so. Then, we both like words, conversation, and listening to conversation. We like to be with bright people and seldom enjoy the company of people who do not speak."

In scientific interactions, fear of hurting one's priority by revealing too much information too early on the one hand, and remaining informed plus claiming priority, on the other, are competing with each other. Strong groups in scientific centers are in better position than small groups in remote places in assuring their priority claims yet participating in information exchange. Watson and Crick worked on the DNA structure in the best possible location. Not only were they surrounded by excellent colleagues, but a stream of visitors kept descending on Cambridge eager to tell them about their latest results. Thus, Peter Pauling reported about his father's progress in his studies of the DNA structure; Chargaff explained to Watson and Crick his discovery of the base ratios; and Caltech-trained Jerry Donohue enlightened them about the preferred forms of the bases in DNA.

Watson and Crick were able to bring together the threads from different research directions and different laboratories that were relevant to their work, and benefited from them. Their target — the DNA structure — was important yet doable. It is possibly the greatest challenge in basic research to find problems that are important yet possible to solve. Rutherford warned his associates never to attempt a difficult problem, meaning to attempt problems for whose solutions the means were available or could be created. Albert Szent-Györgyi hesitated, what problem to attack next after his Nobel Prize. His primary interest would have been the brain, but he opted for muscle research instead. He judged it doable whereas he considered brain research, at a level he would have liked to do it, premature at the time. Crick had a similar dilemma, choosing between the living (brain, the

Francis Crick and James D. Watson at CSHL (courtesy of MRC LMB Archives).

nervous system) and non-living (biopolymers), and chose the latter. He returned to the former at a later stage of his career. Watson's choice of the structure of DNA was an exception in that it was not dictated by any rationale about the doable, especially given his background. It was an outlandish choice; a choice of genius. Partly, this was due to Watson's ignorance about the techniques of X-ray crystallography.

Rita Levi-Montalcini could have had Watson in mind as she contemplated the key to success in science. She wrote in her autobiographical book *In Praise of Imperfection* that "in scientific research, neither the degree of one's intelligence nor the ability to carry out one's tasks with thoroughness and precision are factors essential to personal success and fulfillment. More important for the attaining of both ends are total dedication and a tendency *to underestimate difficulties*, which cause one to tackle problems that other, more critical and acute persons instead opt to avoid."[211] (italics added)

---

[211] Levi-Montalcini, R., *In Praise of Imperfection: My Life and Work*. Basic Books, New York, 1988, p. 5.

Watson and Crick overlooked several key findings by others that could have helped them and they managed to solve their problem in spite of these misses. They could have paid more attention to complementariness right from the beginning had they been familiar with a 1940 paper by Pauling and Delbrück.[212] They wrote it in response to a series of papers by Pascal Jordan who had suggested that interactions between identical or nearly identical molecules or molecular parts would be preferential in constructing stable structures. The problem came up in connection with molecular replication in the cell. Pauling and Delbrück suggested instead that interactions between complementary parts would be more advantageous energetically. Apparently, Watson and Crick were not familiar with their reasoning; Delbrück did not turn their attention to it; and Pauling himself failed to benefit from this work, which he apparently forgot, before publishing a triple helix structure for DNA. In hindsight, the double helix structure with the base-base linkages is a beautiful example of the importance of complementariness in molecular replication. Consideration of Chargaff's base-to-base ratios could have accelerated the discovery of the double helix although it did help in concluding the work. Curiously, Pauling, while familiar with Chargaff's findings, also ignored them, perhaps because of the personal animosity he felt towards Chargaff. We have seen that Watson did not give much importance to the significance of C_2 symmetry unambiguously yielded by the X-ray patterns, although Crick recognized its importance. Franklin did not consider it to be a decisive factor either, at least not initially.

The determination of the structure of the substance of heredity was a great achievement. Considering it from the point of view of structural chemistry, the determination of other biologically important substances were similarly exceptional feats. But there is a tremendous difference; no three-dimensional structure of any biologically important macromolecule has revealed so readily the function of the molecule as the double helix did for DNA. Watson and

[212] Pauling, L., Delbrück, M., "The Nature of the Intermolecular Forces Operative in Biological Processes." *Science* 1940, 92, pp. 77–79.

Crick's discovery of the double helix was a triumph not only of the Cambridge structural school. It was also that of Watson's home base, Delbrück and Luria's phage school, which viewed molecular biology primarily as a quest for biological information and information transfer. Many authors have written about the difference between the British and American approaches,[213] the former giving pre-eminence to structure and the latter to information. Linus Pauling was an exception in the United States, concentrating on structure. Watson managed to bridge the two schools, the American phage group and the British crystallographers although at the time he may have not been aware of this feature of his and Crick's feat. What Watson and Crick were aware of as soon as they had the double helix model was that the structure of DNA "qualified it to convey a genetic message."[214]

The scientific importance of the double helix discovery was augmented and enhanced by the visual appeal of the structure, both by its aesthetic value and educational worth. No wonder it has become a favorite subject for artistic expression.

There is just one last question to address here — although it has been touched upon repeatedly in this writing — and that is Watson's place among the greats in science. Whenever I hear Watson (or Crick for that matter) being mentioned as a Nobel laureate to stress his importance, I feel dissatisfaction.[215] The discovery of Watson and Crick is so much above that level. Similarly, for example, Mendeleev's authority — for the discovery of the Periodic Table of the Elements — would have hardly been enhanced by a Nobel Prize, which had been debated about during his last years, but never accorded to him. That said, however, there is also a funny feeling when Watson and Crick

---

[213] The first may have been J. Kendrew with his review of the book *Phage and the Origins of Molecular Biology* in *Scientific American* 1967, 216, No. 3, pp. 141–144.

[214] Medawar, P., *The Threat and the Glory: Reflections on Science and Scientists*. Oxford University Press, 1991, p. 87.

[215] George Klein tells the story (Klein, G., *The Atheist and the Holy City: Encounters and Reflections*. MIT Press, Cambridge, Massachusetts, 1990, p. 158): Once, when Francis Crick was introduced as having a Nobel Prize, he corrected the presenter saying that he had *the* Nobel Prize.

are being compared with Darwin and Einstein or with Copernicus and Rutherford.

Judging relative greatness among scientists is an intriguing pastime for some. At one point Isaac Asimov decided to compile a list of the ten greatest scientists in history.[216] This was after he had produced his *Biographical Encyclopedia of Science and Technology*,[217] so he was quite qualified for the task, which nonetheless proved to be difficult. He ended up with Archimedes, Darwin, Einstein, Faraday, Galileo, Lavoisier, Maxwell, Newton, Pasteur, and Rutherford (he did not dare going beyond alphabetical order). With such lists the problem is not who gets onto it, but who is left out. Asimov compiled this list in 1963 and ten years later he still did not see any reason to modify it. He made his selection from a larger pool of nominees, which he compiled from 72 names. Of the 72, at the time of Asimov'c compilation, four were alive; Louis de Broglie, Francis Crick, Werner Heisenberg, and Linus Pauling.

Thus, Crick was among Asimov's nominees and Watson was not. Rutherford's name appearing on the list of ten and Crick's among the 72 reminds us of Watson's words that "some day he [Crick] may be considered in the category of Rutherford or Bohr."[218] Asimov was a science writer, not a scientist, but he was also a perceptive observer, and his opinion should not be dismissed lightly. Another compilation, about 100 scientists who changed the world,[219] included Watson, but not Crick. Watson always considered it important how he was being perceived by the world. Suffice it to recall his cultivating a certain image of himself as a peculiar scientist. It was not so much that he often appeared with untied shoelaces in respectable gatherings, but that he untied his shoelaces when arriving at such gatherings.[220]

---

[216] Asimov, I., *Asimov on Chemistry*. Macdonald and Jane's, London, 1975. Chapter 17, "The Isaac Winners," pp. 238–255.

[217] Asimov, I., *Biographical Encyclopedia of Science and Technology*. Doubleday, 1964.

[218] Watson, *Double Helix*, p. 15.

[219] Balchin, J., *Science: 100 Scientists Who Changed the World*. Enchanted Lion Books, New York, 2003.

[220] Seymour Benzer referred to Watson's habit when Benzer presented Watson with a gift on the occasion of the arrival of the Watsons' first child. The present was wrapped in a box in which there was a pair of tiny sneakers for the baby, with untied shoelaces. See, Weiner, *Time, Love, Memory*, p. 65.

Delbrück compared him to Einstein and called him the Einstein of biology, but made no similar statement about Crick. Of course, Delbrück was Watson's one time mentor. Watson himself justified his meticulously recording the events of his life because people in a hundred years might be interested in them.

Only a couple of weeks before my first Watson interview in the spring of 2000, I recorded a conversation with Kenneth G. Wilson, the Nobel laureate physicist. I brought up the question whether Watson and Crick could be counted among the greatest scientists along with Copernicus, Newton, and Einstein? Wilson was laughing as I was asking this question, and I had to ask him whether his reaction was indicative of his under-appreciating the importance of the double helix? Wilson retorted with his own question, "You're a chemist, I'm a physicist. Do the biologists laugh? You should try that. I'll answer your question after you've tried it on some biologists."

The double helix was an epoch making discovery and a most beautiful one at that. Although the intellectual achievement in it would be difficult to compare to Newton's or Kepler's contribution to physics or Darwin's to biology, to Planck's introduction of the quantum, or to Einstein's theory of relativity, there is no doubt about its importance. The former long-time editor of *Nature*, John Maddox ranked it with "Copernicus's successful advocacy of the heliocentric hypothesis."[221] Watson himself stressed that "our finding of the double helix did not represent a particularly difficult intellectual effort."[222] Also, it was to a much greater extent built on the achievements of others than the other contributions mentioned above. Mahlon Hoagland, himself a noted scientist and former professor at Harvard Medical School, writes in his book *Discovery*[223] that the double helix and Darwin's *Origin of Species* "have often been equated in terms of their impact on both science and society." However, in 1981,

[221] Maddox, J., *What Remains To Be Discovered: Mapping the Secrets of the Universe, the Origins of Life, and the Future of the Human Race.* Macmillan, London, 1998, p. 20.
[222] Watson, J. D., "Afterword: Five Days in Berlin." In *Murderous Science: Elimination by Scientific Selection of Jews, Gypsies, and Others in Germany, 1933–1945*, ed. Müller-Hill, B. Cold Spring Harbor Laboratory Press, New York, 1998, p. 193.
[223] Hoagland, M., *Discovery: The Search for DNA's Secrets.* Houghton Mifflin, Boston, 1981, p. 99.

Hoagland found it too early to make such a judgment though he readily granted that the double helix was the biggest event in biology since Darwin.

Watson himself stated semi-modestly that "I and my friends were present at the birth of the DNA paradigm — by any standard one of the great moments in the history of science, if not of the human species."[224] On another occasion, the moderate tone is gone when Watson claims that "Francis Crick and I had given the world the double helix.[225] However, Crick said that "Rather than believe that Watson and Crick made the DNA structure, I would rather stress that the structure made Watson and Crick."[226] It is also true that Rosalind Franklin might have arrived at the solution had Watson and Crick not made the discovery. But, as Aaron Klug pointed it out with insight, in Franklin's hands, the structure would have come out in steps rather than in a splash as it did in Watson and Crick's one-page report. The impact would have been slower and more gradual although, eventually, it would have led to the same consequences. Watson's most original contribution may have been pairing the bases on February 28, 1953. However, that discovery was becoming inevitable as ideas about the structure were accumulating with Chargaff's finding the one-to-one ratio of purine and pyrimidine bases on the one hand and the twofold (C_2) symmetry of the structure strongly suggesting its complementary nature, on the other.

It is also true that for Watson, the C_2 symmetry of the structure was not as revealing as it was for Crick. Back in 1951, he wrote to Delbrück, "Our method is to completely ignore the X-ray evidence."[227] In February 2004, Crick also noted that Watson did not understand the significance of C_2 symmetry of the DNA structure.[228] In contrast, in his work with Donald Caspar on the structure of the

[224] Watson, *Genes, Girls and Gamow*, p. xxiv.
[225] Ibid., p. 5.
[226] Crick, F., "The Double Helix: A Personal View." *Nature* 1974, April 26, pp. 766–771.
[227] Quoted from a letter by Watson to Delbrück, December 9, 1951, in McElheny, *Watson and DNA*, p. 42.
[228] Conversation with Francis Crick in La Jolla, February 7, 2004; See, in Hargittai, *Candid Science VI*, pp. 2–19.

tobacco mosaic virus (TMV) RNA, in 1955, Watson recognized the importance of symmetry arguments.[229]

In his book about the double helix discovery, *What Mad Pursuit*, Crick noted that the discovery of base-pairing was more serendipity than the result of logical thinking. The logical approach according to him would have been to consider Chargaff's rules and then "look for the dyadic [twofold] symmetry suggested by the C_2 space group shown by the fiber patterns. This would have led to the correct base pairs in a very short time."[230] In our conversation, Crick added that Rosalind Franklin did not quite recognize the significance of C_2 symmetry either in solving the DNA structure. Although Franklin was more of a crystallographer than Watson, she had never solved a structure before. She did not have extensive experience in structure analysis and even less with organic systems and polymeric molecules at that. Of course, nobody else had much experience with solving organic polymeric structures either, at that time.

If we define genius as the one who recognizes interconnections between things that others do not recognize, Watson undoubtedly was a genius in the double helix story. Not so much for his role in the discovery itself but in the recognition of its importance as well as in the recognition of its feasibility. It may well be that without Crick, Watson might have not solved the DNA structure. However, without Watson, Crick might have not even addressed himself to this problem. He was working in Perutz's group, ostensibly under Perutz, and his project was related to the protein structure of hemoglobin. He made excursions to other projects in that he tried to solve everybody else's problems (according to W. Lawrence Bragg, he had the habit of "doing someone else's crossword"),[231] which was an indication of his ubiquitous interest, but also an indication of his being bored by his own project. Once Watson arrived, the two threw themselves into the DNA problem and were driven by its overall importance.

[229] Watson, *Genes, Girls and Gamow*, p. 170.
[230] Crick, *What Mad Pursuit*, pp. 65–66.
[231] Hunter, *Light Is a Messenger*, p. 187.

At the same time, DNA was just another important structure to solve for Rosalind Franklin and so it was for Linus Pauling. It was not long before that Pauling had discovered the alpha helix structure of proteins, another helical structure of a biological polymer kept together by intramolecular hydrogen bonding. Pauling spent about 15 years on and off on the problem; there was no urgency in the project almost as if nobody else could have solved it. This was just so because an eminent British group was also studying the same structures and had published a host of wrong models for it, as we have seen above.

Richard E. Marsh, a staff scientist at the California Institute of Technology witnessed and admired Linus Pauling's activities for decades. He worked in X-ray crystallography at Caltech from 1950, and in his words in 2000, "I never felt that Pauling was as interested in the DNA problem as he was in proteins."[232] According to Marsh, Pauling and Corey were disappointed when their proposed structure for DNA was shown to be incorrect. On the other hand, they took pride in learning that one of their students — Jerry Donohue — was helpful in steering Watson and Crick toward the correct structure.

Gunther Stent has narrated vividly, at least on two occasions in print, how Watson recognized that he should solve the DNA structure as his next project. His story turned out to be faulty in facts (see, below), though not in spirit. Watson and Stent were postdocs together in Copenhagen in 1950. Stent described the deciding moment for Watson's interest turning to the three-dimensional structure of DNA in two closely related versions. The essence is given in his essay prepared for the volume commemorating the 50th anniversary of the double helix discovery entitled *Inspiring Science*.[233] According to

---

[232] Hargittai, I., Conversation with Dr. Richard E. Marsh, California Institute of Technology, April 13, 2000, unpublished records. Incidentally, Marsh learned X-ray crystallography at Tulane University in New Orleans from Rose C. L. Mooney of Sophie Newcomb College — the female affiliate of all-male Tulane in 1946. Mooney had created something of a sensation at Caltech when she had been accepted as a graduate student. As R. C. L. Mooney, she was assumed to be a male. When Caltech found out otherwise, they quickly made her a Research Assistant for Pauling instead of a graduate student.
[233] Inglis, J. R., Sambrook, J., Witkowski, J. A. (eds.), *Inspiring Science: Jim Watson and the Age of DNA*. Cold Spring Harbor Laboratory Press, Cold Spring Harbor, New York, 2003.

Niels Bohr (standing at the left), Gunther Stent (standing fifth from the right), James Watson (standing third from the right), Herman Kalckar (standing between Watson and Stent), and Élie Wollman (squatting on the right) (courtesy of Gunther S. Stent).

Stent,[234] Sir Lawrence Bragg gave a lecture at the House of the Royal Danish Science Society in the late fall of 1950. Bragg talked about Linus Pauling's latest discovery, the alpha-helix structure of proteins about which he had just heard from a letter to him by Pauling. Stent calls it "a spirit of true nobility" that Bragg talked about Pauling's much more exciting results rather than the slow progress being made in the search for protein structures by Perutz and Kendrew at the time. Stent "noticed that as Jim was listening to Bragg, he became more and more agitated. After Bragg had finished, Jim turned to me and said, 'THAT'S what we've got to do, Gunther! Get the 3-D structure of DNA, instead of farting around with phage DNA metabolism!'"[235] In January 2003, Stent repeated the story to me in yet greater detail.[236]

---

[234] Ibid., p. 12.
[235] Ibid.
[236] Hargittai, *Candid Science V*, pp. 497–498, p. 500. (Gunther S. Stent)

Stent's story — having now been documented in two different books — sounds realistic, alas, it was a product of Stent's imagination in the way he described it.[237] I have not been able to corroborate that there was such a Bragg lecture in Copenhagen in the fall of 1950. Watson remembers that Bragg's talk occurred after he had read the Pauling paper in the *Proceedings of the National Academy of Sciences of the USA* and he placed it to early summer of 1951.[238] Apparently, he heard about Pauling's alpha-helix after he had returned to Copenhagen from Naples and not before.[239] Graeme K. Hunter, the Canadian biomedical scientist and science historian wrote a book about the life and science of W. Lawrence Bragg, *Light Is a Messenger*.[240] There is no mention of Bragg's visit to Copenhagen in the book. At my request, Hunter looked into the correspondence between Pauling and Bragg at the archives at Oregon State University in Corvallis and at the Royal Institution in London. He found no letter from Pauling to Bragg in the fall of 1950.[241] On the other hand, Pauling wrote to David Harker in March 1951 that he had not mentioned his ideas about protein structures to others, and specifically not to Perutz and Bragg.[242] There was a race between Pauling and the Cambridge group in which Pauling's greater knowledge of structural chemistry prevailed.[243]

Pauling and Corey published a short paper in November 1950 about their alpha-helix and gamma-helix structures,[244] and came out with their detailed report only the next year.[245] This detailed report,

---

[237] Stent came to realize this in an e-mail message to me in September 2006.

[238] J. D. Watson's letter to the author dated September 2, 2003.

[239] Watson, J. D. (with Brown, A.), *DNA: The Secret of Life*. William Heinemann, London, 2003, p. 43.

[240] Hunter, *Light Is a Messenger*.

[241] Private communication from Graeme K. Hunter by e-mail, January 15, 2006.

[242] Hunter, *Light Is a Messenger*, p. 183.

[243] Hargittai, I., Hargittai, M., *In Our Own Image: Personal Symmetry in Discovery*. Kluwer/Plenum, New York, 2000, pp. 88–93.

[244] Pauling, L., Corey, R. B., "Two Hydrogen-Bonded Spiral Configurations of the Polypeptide Chain." *J. Am. Chem. Soc.* 1950, 72, p. 5349.

[245] Pauling, L., Corey, R. B., Branson, H. R., "The Structure of Proteins: Two Hydrogen-Bonded Helical Configurations of the Polypeptide Chain." *Proc. Natl. Acad. Sci. USA* 1951, 37, pp. 205–211.

which was part of a series of papers from Pauling's laboratory, made Perutz realize that Pauling got the right structure and the Cambridge group brought out erroneous models.[246]

Watson's account, *The Double Helix* makes no mention of any Bragg lecture and, accordingly, no impact of it on him. He describes his meeting with Maurice Wilkins in Naples in the spring of 1951, and says that it was the first time that he became excited about X-ray work on DNA.[247] His description of their Naples encounter is consistent with Wilkins's account.[248] At my request, the Archives of the Niels Bohr Library in Copenhagen made a search of any indication of Bragg's possible visit to Copenhagen. According to their records, there was an exchange of letters between Bohr and Bragg in the spring of 1951, according to which Bragg was to visit Copenhagen in June 1951.[249] This corroborates Watson's memory about attending a Bragg lecture some time after his crucial meeting with Wilkins earlier in 1951 in Naples. In 2006, I once again asked Watson about this story, and his answer was unambiguous: "In contrast to Gunther Stent's memory, I long have placed Bragg's visit to Copenhagen as occurring in the summer of 1951, two months after my seeing Wilkins's DNA X-ray photo. In any event, it was the glimpse of Wilkins's photo that made me want to move to an X-ray crystallographic lab devoted to macromolecules."[250]

---

[246] Hargittai, *Candid Science II*, pp. 286–288 (Max F. Perutz); See, also Perutz, M. F., *I Wish I'd Made You Angry Earlier.* Oxford University Press, 1998, pp. 173–175.

[247] Watson, *Double Helix*, p. 22.

[248] Wilkins, *Third Man*, pp. 138–139.

[249] Private communication from Felicity Pors of the Niels Bohr Archive, Copenhagen, to the author, dated August 4, 2006. The letter in the microfilmed Bohr Scientific Correspondence (AHQP; BSC, mf.27) from Niels Bohr to W. Lawrence Bragg is dated May 31, 1951. In it, Bohr thanks for a letter from Bragg of May 7, 1951, with information about Bragg's planned visit to Copenhagen (this letter has not been found). Bohr writes to Bragg that he will be in Glasgow in June, but hopes to be back on Copenhagen on Saturday, June 23, to have a chance to see Bragg before Bragg and his wife leave for Stockholm.

[250] J. D. Watson's letter to the author dated June 20, 2006.

Although Stent's narrative about the timing of Bragg's lecture and its impact on Watson did not bear out scrutiny, he was a close observer of Watson. He remembered their first meeting:[251]

When I first met Jim at Cold Spring Harbor in 1948, I was surprised when he told me that he was from Chicago's South Side (where I too was from), because he didn't talk like a Southsider. He spoke with a weird accent. I thought that, maybe, he was a foreigner who had learned English in a Berlitz school. It turned out that Jim was imitating Roger Stanier, the brilliant Canadian microbiologist whom he had met at Indiana University. Stanier was born on Vancouver Island, off the Coast of British Columbia, where they speak a peculiar, old-fashioned kind of English.

Did Jim imitate Delbrück at some point?
Not in his speech, but in Max's Berlin-style short haircut and some mannerisms. By the time of his discovery of the DNA double helix, Jim had changed his hair styling to the long, Cambridge-style locks.

According to Watson,[252] "...imitation is very important. Even though imitation might initially sound the cheap way to go, it is only a question of whom you imitate. If you are trying to mimic someone very good, you really can't pull it off, but what comes out might nevertheless be fairly interesting."

Returning to the question of Watson's place among the greatest scientists, to be a genius is not the same as being a great scientist. There is no objective measure for genius, and there is none for greatness in science either. A great scientist may be one who opens up new vistas in science even if this scientist makes no significant specific discoveries. On the other hand, great discoverers are not necessarily great scientists. There are many one-discovery scientists who make a

[251] Hargittai, *Candid Science V*, p. 497. (Gunther S. Stent).
[252] Watson, *A Passion for DNA*, p. 117.

truly important breakthrough in science, but accomplish never again anything of real significance, and soon after the seminal discovery disappear in oblivion. In spite of the tremendous publicity of the double helix, overshadowing many other discoveries, Watson cannot be considered a one-discovery scientist. One of the remarkable characteristics of his career is that he has maintained his presence at the frontiers of science for a long time although he never did in direct research anything even remotely comparable to the discovery of the double helix again.

Two years after my conversation with Kenneth Wilson, referred to above, we took up the question about Watson's greatness again.[253] Wilson recognized the absolutely central role of DNA and its double helix structure in genetics. However, he would place the founders of quantum mechanics ahead of Watson and Crick "in terms of both the originality and creativity of their work and the breadth of their impact on our understanding of the world that we live in." Wilson also called attention to the fact that today there are many more scientists working than in the times of Copernicus and Newton or even of Darwin. He thinks that

> "the dominant story in science is not the contributions of individuals, not even of Einstein, let alone Watson and Crick. Instead the BIG story (at least for me) is the growth of institutions that are dependent on the research of increasingly large numbers of scientists... The areas of the economy that have been transformed include medicine, transportation, telecommunications and information storage, energy, industrial materials, defense, large-scale construction (including skyscrapers and bridges), and media and entertainment. All of these transformations have been accomplished with the crucial help of a growing base of knowledge: scientific knowledge, engineering knowledge, and professional knowledge in areas such as medicine and law. In fact, knowledge in toto has become the dominant source for economic wealth, social

---

[253] Hargittai, *Candid Science IV*, pp. 544–545. (Kenneth G. Wilson).

progress (if any), and military might ... as I learn more about the growth of knowledge as a whole, I become less inclined to take any list of 'big names' in science seriously. I admire the accomplishments of Newton and Einstein, and of Darwin and Watson and Crick too. But I give equal weight to seemingly far more mundane matters such as the growth of reference materials (encyclopedias, dictionaries, and the like) in libraries, without which hardly anyone would have a possibility of learning about the accomplishments of the big and not-so-big names in science...."

Comparison of greatness among scientists may both be futile and instructive; futile because by contrasting entirely different activities, there are no criteria to make an objective judgment; at the same time, it may be instructive, because the individuals involved are important players in human history, and even we — ordinary people — may have a say about them in such deliberations. Similarly futile, but not useless either may be the discussions aiming at choosing the most fundamental scientific discoveries. Futile for similar reasons as judging relative greatness; at the same time, such discussions, again, facilitate our understanding of the specific discoveries. Important scientists also get involved in such discussions.[254] Thus, countering the often repeated claim that elementary particle physics is more fundamental than other branches of science, the Nobel laureate physicist, Philip Anderson maintains that it is not more fundamental than, say, "what Francis Crick and James Watson did in discovering the secret of life."[255] Steven Weinberg tries to provide a finer formulation about the importance of not so much of Crick and Watson's discovery than that of DNA itself. He says, "It is not that the *discovery* of DNA was fundamental to all of the *science* of life, but rather that DNA itself is fundamental to all life itself" (italics by Weinberg).[256] In our discussion here, we are more concerned with the importance of specific scientists as individuals than with science in general.

---

[254] Weinberg, S., *Dreams of a Final Theory*. Vintage Books, New York, 1993, pp. 55–57.

[255] Ibid., p. 55, Weinberg quoting Philip Anderson's letter to *The New York Times*, June 8, 1986.

[256] Ibid., p. 57.

To many of the famous biomedical scientists I have recorded conversations with I posed the question, whom they would consider to be their hero? Many named Einstein among the dead, and many named Crick among the living (Crick was still alive during the bulk of my interviews). None named Watson. However, greatness is not a question of popularity, and at this point I would like to offer my own view about this question. The careers of both Crick and Watson stemmed from their discovery of the double helix structure of DNA. Without this discovery we can only speculate about Crick's and Watson's possible paths. Crick would have finished his thesis work at the Cavendish Laboratory sooner or later, and might have gone on to a solid career in science in Great Britain, making important but fragmentary contributions, fertilizing the research of others with his critical and forceful approach, and might have ended up with no single very outstanding discovery. Watson was more determined to have a distinguished career and was capable of very concentrated and sustained efforts as his textbook writing demonstrated in the 1960s. He might have ended up in a revered position in American academia, though not necessarily in a Harvard chair.

The discovery of the double helix catapulted both Watson and Crick into a unique orbit, and the question might be asked whether there was any decisive difference between the two in this discovery? The specific contributions of Watson and Crick to the discovery of the DNA structure are impossible to discern. Although the concept of base pairing that proved to be so decisive in the completion of the discovery might be assigned more to Watson than Crick, it was still largely a result of their interactions. If Watson had not matched the bases when he did they might have come up with it a day or a week later in some other sequence, and so on. So there is no way to assess their relative contributions in the discovery. What remains to examine is whether there is any discernible difference between them in placing themselves onto the path of working on the DNA structure. Here we do recognize a difference. Had Watson not come to Cambridge and started to talk with Crick, Crick might have well continued his thesis work on protein structure. Crick might have thought of the importance of nucleic acids — and Watson stated in

The Double Helix[257] that he had — and might have thought that their structure should be determined, but the fact remains that he was working on a protein structure and there is no indication that he might have switched to nucleic acids any time soon. In 1946, Crick advised Maurice Wilkins to "get himself a good protein" to study.[258] As late as 1951, but still before Watson's arrival, Crick did not appear interested in what Wilkins had to say about a possibly helical DNA structure at a seminar in Cambridge organized by Max Perutz.[259] Yet — according to Crick's biographer — something may have changed between 1946 and 1951 in Crick's thinking as he began ascribing more importance to DNA than before.[260] Watson's determination on working on DNA structure has been amply evidenced. He went to Cambridge with the DNA structure on his mind, and he forfeited his scholarship in his stubborn determination. This was a decisive difference, which — in my view — tilts the judgment in Watson's favor.

Watson's public persona has outshined that of Crick's, but knowing that he cultivated his fame, whereas Crick did nothing to this end, it remains a question whether this difference would be sustained in the long run? Cold Spring Harbor Laboratory where Watson reigns — regardless of whether he is director, president, or chancellor — is not only an excellent institution of biomedical research; it is also Watson's shrine. This comes across whether we observe his larger than life painting in Grace Auditorium, hear about the Watson Graduate School, marvel at the double helix sculptures in the foyer of Grace Auditorium and on the hamlet near the Watsons' home, or admire the Hazen Tower in front of the Beckman Laboratory, with the initials of the four bases carved one each on its four sides. The entire Cold Spring Harbor Laboratory is a testimony of his impact, and the fate of the Laboratory will impact Watson's fame in the years to come.

---

[257] Watson, *Double Helix,* p. 31.
[258] Ridley, M., *Francis Crick: Discoverer of the Genetic Code.* Atlas Books, Harper Collins Publishers, New York, 2006, p. 25.
[259] Ibid., p. 51.
[260] Ibid., p. 32.

Left: Painting of J. D. Watson by Lewis Miller in Grace Auditorium at Cold Spring Harbor Laboratory; right: Double Helix sculpture by Bror Marklund at Uppsala University (photos by the author).

Watson has masterfully built up a tremendous image. He succeeded in this more than others because his self-aggrandizement was combined with outward shyness and the absence of oratory abilities. It was also combined with outstanding science. The single most important feature of Watson's image building has been his becoming increasingly identified with DNA, and not just its structure, but the substance itself. His life "has encompassed the Age of DNA."[261]

The fame of individual scientists may fade, but DNA is eternal.

[261] Pollack, R., *Signs of Life: The Language and Meanings of DNA*. Houghton Mifflin Co., Boston and New York, 1994, p. 4.

Sampler of Quotable Watson

Watson, J. D., *The Double Helix: A Personal Account of the Discovery of the Structure of DNA.* A Norton Critical Edition, Norton, New York, 1980.

p. 3.

...styles of scientific research vary almost as much as human personalities.

Watson, J. D., *A Passion for DNA: Genes, Genomes and Society.* Oxford University Press, © 2000 by James D. Watson.

p. 5.

Offending somebody was always preferable to avoiding the truth... [*Values from a Chicago Upbringing*, 1993]

p. 93.

...we often take joy in the discoveries of others, but often only in proportion to the extent that we were not close to the same objective. [*The Dissemination of Unpublished Information*, 1973]

p. 225.

Genetic disease is the price we pay for the extraordinary evolutionary process that has given rise to the wonders of life on Earth. [*Good Gene, Bad Gene: What Is the Right Way to Fight the Tragedy of Genetic Disease?*, 1997]

Watson, J. D., Tooze, J., ***The DNA Story: A Documentary History of Gene Cloning.*** W. H. Freeman and Co., San Francisco, 1981. © 1981 by James D. Watson and John Tooze.

p. 160.
You've got to smell a little smoke before you decide where the fire is. [Watson, J., "Remarks on Recombinant DNA." *CoEvolution Quarterly*, Summer 1997, 14, p. 40]

Tibor Szántó's interview with James D. Watson at the Cold Spring Harbor Laboratory, on September 24 and 27, 1986.
A sentimental science would not survive.

James D. Watson quoted in **Judson, H. F., *The Eighth Day of Creation: Makers of the Revolution in Biology*,** Expanded Edition. Cold Spring Harbor Laboratory Press, Cold Spring Harbor, New York, 1996.

p. 4.
"It's necessary to be slightly underemployed if you are to do something significant." [from Judson's interviews with Watson, 1970–1971]

Watson, J. D., **"Looking forward."** *Gene* 1993, 135, pp. 309–315.

p. 315.
It is generally much cheaper to prevent most diseases than to cure them.

p. 315.
What genetic procedures we regard as good or evil may vary from one moment of human existence to another.

Watson, J. D., **"The Human Genome Project: Past, Present, and Future."** *Science* 1990, 248, pp. 44–49.

p. 44.
... the Human Genome Project... A more important set of instruction books will never be found by human beings.

p. 48.
The nations of the world must see that the human genome belongs to the world's people, as opposed to its nations.

Watson, J. D., "A Personal View of the Project." In *The Code of Codes: Scientific and Social Issues in the Human Genome Project,* eds. **Kevles, D. J., Hood, L.** Harvard University Press, Cambridge, Massachusetts, and London, England, 1992, pp. 164–173.

p. 164.
I have spent my career trying to get a chemical explanation for life, the explanation of why we are human beings and not monkeys. The reason, of course, is our DNA. If you can study life from the level of DNA, you have a real explanation for its processes.

James D. Watson quoted in **Weiner, J.,** *Time, Love, Memory: A Great Biologist and His Quest for the Origins of Behavior.* Alfred A. Knopf, New York, 1999. [This book is about Seymour Benzer]

p. 175.
We used to think our fate was in the stars. Now we know, in large measure, our fate is in our genes. [quoted from James D. Watson in *Time* magazine in 1989]

James D. Watson quoted in **Appleyard, B.,** *Brave New Worlds: Staying Human in the Genetic Future.* Viking, New York, 1998.

p. 150.
The end result of the human genome project on society will finally be to make people realize we are the products of evolution, not

of message from the sky. Finally they are going to find it impossible to ignore. [from a conversation of Bryan Appleyard with Watson]

Watson, J. D., **"The Pursuit of Happiness: Liberty Medal Address, City of Philadelphia, July 4, 2000."** *The Chemical Intelligencer*, October 2000, 6(4), pp. 47–48.

p. 48.
...it is discontent with the present that leads clever minds to extend the frontiers of human imagination.

Watson, J. D., quoted in **Cook, M., *Faces of Science.*** W. W. Norton & Co., New York, London, 2005.

p. 160.
I am a scientist in large part because I was born highly curious.

James D. Watson through the Eyes of Others: A Sampler

Salvador Luria on Watson

Luria, S.E., *A Slot Machine, A Broken Test Tube: An Autobiography.* Harper & Row, New York, 1984.

p. 89.
Like an arrow released from a crossbow or a prisoner paroled from Alcatraz, Jim rushed from Copenhagen to Cambridge and settled there, without awaiting the approval of the August committee that had awarded him his two-year fellowship.

p. 42.
Watson later developed into a shrewd administrator (he is now head of the Cold Spring Harbor Laboratory) and a rambunctious statesman of science.

George W. Beadle quoted in Berg, P., Singer, M., *George Beadle, An Uncommon Farmer: The Emergence of Genetics in the 20th Century.* Cold Spring Harbor Laboratory Press, Cold Spring Harbor, New York, 2003.

p. 217.
Because there are literally no other James Watsons in the world, he is absolutely essential to our research work... He is simply irreplaceable.

[in connection with obtaining deferrment from military service for Watson]

George Beadle on Watson and Crick

Beadle, G., Beadle, M., *The Language of Life: An Introduction to the Science of Genetics.* Victor Gollancz, London, 1966.

pp. 166–167.
Watson and Crick did no research — as laymen understand the word. They reread all the literature about DNA; they covered paper and blackboards with formulas and equations; they snipped "nucleotides" from thin metal sheets and used them for model-making. But mostly they just *thought.* [italics in the original]

W. Lawrence Bragg on Watson and Crick

In a letter of reference on Watson by Bragg to John Raper of Harvard University in March 1955, quoted in Hunter, G. K., *Light Is a Messenger: The Life and Science of William Lawrence Bragg.* Oxford University Press, 2004, p. 228.

It was my impression that Watson was responsible for the brilliant and imaginative ideas in this work. He really has, I think, a touch of genius. Crick is a young man of great energy and wide reading who supplied knowledge about stereochemistry and symmetry and about X-ray diffraction which Watson lacked. But it seemed to me that Watson supplied the main girders of the structure. Crick knocked in the rivets which held it together. The trouble was that Crick, who is a very voluble young man, always did all the talking and Watson, being somewhat shy and sensitive, never had a chance to put his own case. I think he felt this very much and it led to certain difficulties between them...He is altogether a most interesting young man who much attracted me.

Francis Crick on Watson

Crick, F., "The Double Helix: A Personal View." *Nature* 1974, April 26, pp. 766–771.

p. 771.
[at the time of the double helix discovery] Watson was regarded, in most circles, as too bright to be really sound.

p. 771.
Jim was always clumsy with his hands. One had only to see him peel an orange... [would-be opening line of a book Crick never wrote with the title, *The Loose Screw*]

Linus Pauling on Watson

Hager, T., *Force of Nature: The Life of Linus Pauling.* Simon & Schuster, New York, 1995.

p. 412 (Hager quoting Pauling from his interview with Pauling)
Pauling remembered Watson as "something of a monomaniac" where DNA was concerned...

Peter Medawar on Watson

Medawar, P., "Lucky Jim." In *Pluto's Republic.* Oxford University Press, 1982, pp. 270–278. (originally it appeared as a review of Watson, J. D., *The Double Helix*, in the *New York Review of Books*, March 28, 1968)

p. 273.
It could be said of Watson that, for a man so cheerfully conscious of matters of priority, he is not very generous to his predecessors. The mention of Astbury is perfunctory and of Avery a little condescending. Fred Griffith is not mentioned at all. Yet a paragraph or two would have done it without derogating at all from the splendour of his own achievement. Why did he not make the effort?

It was not lack of generosity, I suggest, but stark insensibility. These matters belong to science history, and the history of science bores most scientists stiff. A great many creative scientists (I classify Jim Watson among them) take it quite for granted, though they are usually too polite or too ashamed to say so, that an interest in the history of science is a sign of failing or unawakened powers...

p. 274.
I do not think Watson was lucky except in the trite sense in which we are all lucky or unlucky — that there were several branching points in his career at which he might easily have gone off in a direction other than the one he took.

p. 275.
Watson ... in addition to being extremely clever he had something important to be clever *about.* (italics in the original)

p. 275.
Lucky or not, Watson was a highly privileged young man.

Maurice Wilkins on Watson

Wilkins, M., *The Third Man of the Double Helix: The Autobiography of Maurice Wilkins.* Oxford University Press, Oxford, 2003.

pp. 138–139.
[Referring to a small meeting in Naples in the spring of 1951 on the structure of large molecules in the living cell]
"He [Watson] was very excited about my presentation when I showed our photographs that showed that DNA was crystalline, and he decided to work on the structural chemistry of nucleic acids and proteins. ... I did not understand much of what Watson had to say. He talked to me about genes and viruses, but I did not know much about bacteriophages and could not make much sense of what he was telling me. Watson was one among many interesting new people at this exciting conference, and I did not spend much time with him. ... Later on, in calmer environments, I was to find Jim Watson very clear and interesting."

Erwin Chargaff on Watson and Crick

Chargaff, E., *Heraclitean Fire: Sketches from a Life Before Nature*. The Rockefeller University Press, New York, 1978, pp. 101, 102.

The impression: one, thirty-five years old; the looks of a fading racing tout, something out of Hogarth ("The Rake's Progress"); Cruickshank, Daumier; an incessant falsetto, with occasional nuggets glittering in the turbid stream of prattle. The other, quite undeveloped at twenty-three, a grin, more sly than sleepish; saying little, nothing of consequence; a "gawky young figure, so reminiscent of one of the apprentice cobblers out of Nestroy's Lupazivagabundus." I recognized a variety act, with the two partners at that time showing excellent teamwork, although in later years helical duplicity diminished considerably. The repertory, however, unexpected.

…

It was clear to me that I was faced with a novelty: enormous ambition and aggressiveness, coupled with an almost complete ignorance of, and a contempt for, chemistry, that most real of exact sciences — a contempt that was later to have a nefarious influence on the development of "molecular biology." Thinking of the many sweaty years of making preparations of nucleic acids and of the innumerable hours spent on analyzing them, I could not help being baffled...

Max Perutz on Watson and Crick

Perutz, M., *I Wish I'd Made You Angry Earlier: Essays on Science, Scientists, and Humanity*. Oxford University Press, 1998.

p. 188.
They shared the sublime arrogance of men who had rarely met their intellectual equals. Crick was tall, fair, dandyishly dressed, and talked volubly, each phrase in his King's English strongly accented and punctuated by eruptions of jovial laughter that reverberated through the laboratory. To emphasize the contrast, Watson went around like a tramp, making a show of not cleaning his one pair of shoes for an entire term (an eccentricity in those days), and dropped his sporadic

nasal utterances in a low monotone that faded before the end of each sentence and was followed by a snort.

p. 189.
Like Leonardo, Crick and Watson often achieved most when they seemed to be working least.

In Judson, H. F., *The Eighth Day of Creation: Makers of the Revolution in Biology*, Expanded Edition. Cold Spring Harbor Laboratory Press, Cold Spring Harbor, New York, 1996.

p. 4 [Judson quoting Perutz from a conversation on May 29, 1968].
[Watson] never made the mistake of confusing hard work with hard thinking; he always refused to substitute the one for the other. Of course he had time for tennis and girls.

André Lwoff on Watson in his review of *The Double Helix*; *Scientific American*, July 1968, pp. 133–138; reprinted in the Norton Critical Edition of *The Double Helix*, pp. 224–234.

p. 230.
May God protect us from such friends!

p. 231.
Jim has received golden gifts: the aptitude to formulate attack and solve important problems; the power of abstraction from the outer world, the power to "dream" the problems. Intuition and logic are seldom both present in one person at such a high level. The brain functions with remarkable efficiency. Moreover, Jim has risen above his great discovery and continues to work with success. It would appear that these brilliant gifts are not balanced by an equal development of affectivity.

p. 232.
Jim allows himself to be sensitive only insofar as the person involved reflects his own interests.

p. 232.
His characteristics are essentially cold logic, hypersensitivity and lack of affectivity.

pp. 233–234.

Perhaps some day Jim will learn that all impressions, however witty they may seem, are not necessarily suitable for publication, that human beings are easily hurt and that the wounds, particularly to self-esteem, are painful and slow to heal.

André Lwoff quoted in Judson, H. F., *The Eighth Day of Creation: Makers of the Revolution in Biology*, Expanded Edition. Cold Spring Harbor Laboratory Press, Cold Spring Harbor, New York, 1996.

p. 122.

It is evening in the solemn drawing room of the Abbeys. In the room is a 15th-century oak table, on which there is a bust of Henri IV. A young American scientist wearing shorts, has climbed on the table and is squatting beside the king. An unforgettable vision!

François Jacob on Watson and Crick

Jacob, F., *The Statue Within: An Autobiography*. Unwin Hyman Ltd., London, 1988.

p. 264 (based on observing him at an Oxford colloquium in 1952).

...Jim Watson was an amazing character. Tall, gawky, scraggly, he had an inimitable style. Inimitable in his dress: shirttails flying, knees in the air, socks down around his ankles. Inimitable in his bewildered manner, his mannerism: his eyes always bulging, his mouth always open, he uttered short, choppy sentences punctuated by "Ah! Ah!" Inimitable also in his way of entering a room, cocking his head like a rooster looking for the finest hen, to locate the most important scientist present and charging over to his side. A surprising mixture of awkwardness and shrewdness. Of childishness in the things of life and maturity in those of science. In the little world of the phage attending the colloquium, Jim created a sort of revolution when, instead of reading Luria's report, he brandished a letter he had just received from Al Hershey, another American prima donna of the phage chorus. It concerned a new finding. A neat, irrefutable experiment.

p. 270 (about the 1953 meeting in Cold Spring Harbor where the double helix was first reported).

...the star turn was Jim Watson's description of the structure of DNA, which he had just worked out with Francis Crick. ... His manner more dazed than ever, his shirttails flying in the wind, his legs bare, his nose in the air, his eyes wide, underscoring the importance of his words, Jim gave a detailed explanation of the structure of the DNA molecule; breaking into his talk with short exclamations the construction of atomic models to which he devoted himself at Cambridge with Francis Crick; the arguments based on X-ray crystallography and biochemical analysis; the double helix itself, with its physical and chemical characteristics; finally, the consequences for biology, the mechanism that underlay the recognized properties of genetic material: the ability to replicate itself, to mutate, to determine the characters of the individual. For a moment, the room remained silent. There were a few questions. ... But no criticism. No objections. This structure was of such simplicity, such perfection, such harmony, such beauty even, and biological advantages flowed from it with such rigor and clarity, that no one could not believe it untrue....

Renato Dulbecco in *Inspiring Science* (p. 7).

"Jim had few filters between his mental processes and verbal expression of his opinions."

Sydney Brenner on Watson

Brenner, S., *My Life in Science*. As told to Wolpert, L., Friedberg, E. C., Lawrence, E., eds. Biomed Central Limited, London, 2001.

pp. 26–27.

My initial impression of Jim Watson was of this rather eccentric, bright person who didn't pay as much attention to me as I would have liked, and who walked very rapidly with long strides — because we went for a walk. And someone who knew all the important people. You have to realize that this was my first meeting with someone who actually knew Delbrück and Luria. I mean, I knew Luria's work; I was doing Luria-Delbrück experiments. And Jack Dunitz knew Pauling. It was a great thrill to meet people for the first time who knew other people who had experiments named after them.

Sydney Brenner in *Inspiring Science*, p. 69.

"Jim is a very skilled administrator who understands the politics of getting things done. ... He knows that you must reach the hearts and minds of people before you go for their pockets and how important it is to convince people that they are direct participants in a collaboration."

"Jim ... feels that only he can utter the unutterable."

"Worrier and Warrior, Jim has been the guardian of DNA for the past 50 years."

Rita Levi-Montalcini on Watson

Levi-Montalcini, R., *In Praise of Imperfection: My Life and Work.* Basic Books, New York, 1988.

p. 138.
In 1947, Levi-Montalcini visited Salvador Luria at Indiana University in Bloomington. Luria introduced "the most brilliant of his students" James Watson to her and she recorded her impressions of him as follows: "He had the appearance of an adolescent, his brow still shaded with blond hair which was to give way to premature baldness, His absorbed and dreamy look, his slenderness, and his absentminded way of moving reminded me of a famous Picasso Harlequin of the Blue Period. He took no interest in me whatsoever and left immediately with a hasty goodbye. Our brief encounters over the following years, when the unknown adolescent became the famous Watson, were characterized by his absolute indifference toward me — an attitude I saw as part of his well-known anti-feminism. I was never troubled by it."

Robert L. Sinsheimer on Watson and Crick and on Watson

Sinsheimer, R. L., *The Strands of a Life: The Science of DNA and the Art of Education*. University of California Press, Berkeley, etc., 1994.

p. 86.
The elucidation of the double helix structure by Watson and Crick has taken on mythic dimensions. The authors of the myths, non-scientists and (even) scientists like Watson himself, at the time ignorant of the history of DNA research, have presented a sometimes self-serving scenario akin to the primitive myths of creation in which the world — or the DNA structure — is derived from a formless void. Of course, it wasn't like that.

p. 90.
Jim Watson is the stuff of which *People* magazine is made. Brilliant, arrogant, verbally crude, with a skewed, off-center personality, fond of publicity, he sees the world in black and white with little gray in between. In his career, he has consistently demonstrated excellent scientific judgment and has been a superb director of the Cold Spring Harbor Laboratory.

Brenda Maddox's description of Watson and Crick, and contrasting them with Franklin and Wilkins

Maddox, B., *Rosalind Franklin: The Dark Lady of DNA*. Harper Collins, London, 2002.

p. 159.
Affinity is no easier to explain than antipathy. The two men clicked just as much as Rosalind and Maurice repelled each other. Both were irrepressible talkers, with quick minds and complementary expertise. Watson knew biology and genetics; Crick was a physicist who had taught himself X-ray crystallography. The words poured out and continued every day over lunch at the Cavendish's local pub, The Eagle. Both laughed a great deal, even when, as was not always the case, they were discussing how genes might be put together — Crick in a loud bark, Watson in a snuffling snort that showed a lot of his gums. Crick had the merry, knowing eyes of the super-bright; Watson

the horizon-scanning gaze of a radar. Neither had an ounce of depression in him, while Rosalind and Maurice, in their very different ways, were prey to melancholy.

Robert Olby on Watson and Crick

Olby, R., *The Path to the Double Helix: The Discovery of DNA*. Dover Publications, 1994 (originally published by the University of Washington Press, Seattle, 1974).

p. 297.
When Watson and Crick met in 1951 they presented a striking contrast. Crick was a confident, ebullient, articulate Englishman from the middle classes, with a loud laugh, an insatiable curiosity, who could talk for hours on end, and was decidedly extroverted. Watson was diffident in manner, his words were brief, his curiosity was strictly confined to scientific subjects and ornithology. He was introverted, he appeared as a "loner." Whereas Crick at 35 had still not completed his PhD thesis Watson had completed his in 1950 at the age of 22.

p. 297, quoting Paul Weiss on Watson at Indiana University.
He was [or appeared to be] completely indifferent to anything that went on in the class; he never took any notes and yet at the end of the course he came top of the class.

p. 298.
Thus, the quiz kid from Chicago, the bird-watcher, could be seen in Bloomington, a strange figure, always clad casually, usually wearing tennis shoes, tall, thin and awkward-looking. Watson did not have a fund of small talk and he lacked an affable friendship with the other graduate students in his year. With David Nanny (now at Illinois) he did form a close friendship based on mutual respect, but it was characteristic of Watson to seek out older more experienced men and talk with them. Some of his fellow students thought of him as "way-out." He would walk past them on the corridor with faraway look in his eyes. To those he did not wish to talk he could be reserved, almost disdainful. This young aspiring scientist had come

up quickly to graduate at 19. He was used to the companionship of bright lads at the University of Chicago and had no time for numskulls.

Horace F. Judson on Watson

Judson, H. F., *The Eighth Day of Creation: Makers of the Revolution in Biology*, Expanded Edition. Cold Spring Harbor Laboratory Press, Cold Spring Harbor, New York, 1996.

pp. 3–4.
[in a comparison with Max Perutz] Watson is different: impatient, skeptical, forever discontent, swallowing ends of words and ends of thoughts, with a face at 50 only a little less gaunt than in the Cambridge photographs from 1953, with ice-blue, protuberant, even slightly wild eyes and fully modeled lips that retreat in a twitchy preoccupied smile.

p. 11.
Neither Watson nor Crick was a biochemist. They were ignorant of a long and erudite scientific tradition, but at least they were not blinded by it.

p. 139, Judson quoting an unnamed source.
He [Watson] was very clever at making a little bit of knowledge go a very long way.

Philip Morrison on Watson in his review of *The Double Helix; Life*, March 1, 1968, p. 8; reprinted in the Norton Critical Edition of *The Double Helix*, pp. 175–177.

p. 176.
Watson has a sharp eye and honest tongue.

F. X. S. on Watson in his review of *The Double Helix; Encounter*, July 31, 1968, pp. 60–66; reprinted in the Norton Critical Edition of *The Double Helix*, pp. 177–185.

p. 180.
...Watson is a very intricate person.

Richard C. Lewontin on Watson in his review of *The Double Helix; Chicago Sunday Sun-Times*, February 25, 1968, pp. 1–2; reprinted in the Norton Critical Edition of *The Double Helix*, pp. 185–187.

p. 187.
James Watson was consumed with ambition for public praise and approbation, for the highest honor that a doting company of his peers could give.

Howard Green in 1993 (quoted in *Inspiring Science*, p. 111; referring to the June 1953 meeting at CSHL),

"I remember asking him [Watson] something and had the feeling he thought I wasn't worth talking to (quite right)."

Theodore T. Puck in 1993 (quoted in *Inspiring Science*, p. 112; referring to the June 1953 meeting at CSHL),

"... Jim ...was then as he is today — deeply introspective, impatient with sloppy thinking, and interested only in experiments and logical analysis that would open up new vistas in science."

Joseph S. Gots in 1993 (quoted in *Inspiring Science*, p. 113; referring to the June 1953 meeting at CSHL),

"...My impression of young Jim was that he was a bit shy, but highly opinionated and intolerant of lesser intellects."

William F. Dove on Watson

Dove, W. F., "Closing the Circle: A. D. Hershey and Lambda I." In *We Can Sleep Later: Alfred Hershey and the Origins of Molecular Biology*, ed. Stahl, F. W., Cold Spring Harbor Laboratory Press, Cold Spring Harbor, New York, 2000.

p. 46.
"Jim Watson has a talent for selecting the right person for the task."

Shostak, S., *Death of Life: The Legacy of Molecular Biology.* Macmillan, London, 1998.

p. 85.
Watson and Crick did not work in an intellectual vacuum, and their discovery did not arise fully formed as if springing from the head of Zeus (despite suggestions that the hypothesis rose like a Titan missile launched from a nuclear submarine). On the contrary, the discovery of the double helix was possible because, for decades, chemists and physicists had taken an interest in what they considered leading problems in biology...

p. 105.
The contribution of Watson and Crick was not historic because it solved a rather simple problem of structural chemistry.

p. 106.
The genius of Watson and Crick was to couple the structure of a molecule with the concept of the gene.

p. 123.
As happens in young love, the gene swept Watson off his feet and blinded him to anything or anyone else. (Does first love also excuse his misdemeanors and felonies?)

Appleyard, B., *Brave New Worlds: Staying Human in the Genetic Future.* Viking, New York, 1998.

p. 164.

Both Francis Crick and James Watson, for example, believe that part of their task is to rid people of illusions so that they may fully accept the truths of science.

Appendix 3

Watson's The Double Helix *has been in print ever since it appeared. It was also published in the Norton Critical Edition series,* which is rather unique for a contemporary author. This critical edition was edited by Gunther S. Stent and it contains in addition to the original book, commentaries, book reviews, and reprints of original papers. The reprinted reviews are led by Stent's review of reviews. Among the reviews reprinted in the critical edition, Chargaff's review in* Science *is missing because Chargaff declined permission to reprint it. However, Stent discusses it at length in his review of reviews. One review by an important scientist is entirely missing not only from the volume but also from Stent's review of reviews. This is J. Desmond Bernal's review with which Stent was not familiar, perhaps because it appeared in a little known communist periodical,* Labour Monthly.† *Bernal's review is well worth reading as it contains not only an evaluation of the book — appreciative to be sure — but also additional aspects of the double helix story by this scientist, who was a pioneer and major player in the early days of molecular biology. This is why the review is reproduced here in full.*

* Watson, J. D., *The Double Helix: A Personal Account of the Discovery of the Structure of DNA*, ed. Stent, G. S. A Norton Critical Edition, W. W. Norton & Co., New York and London, 1980.
† Bernal, J. D., "The Material Theory of Life." *Labour Monthly* July 1968, pp. 323–326.

The Material Theory of Life

Bernal, J. D., FRS

'No species is ever changed, but each remains so much of itself that every kind of bird displays on its body its own specific markings. This is a further proof that their bodies are composed of changeless matter. For, if the atoms could yield in any way to change, there would be no certainty as to what could arise and what could not, . . . nor could successive generations so regularly repeat the nature, behaviour, habits and movements of their parents.'

Lucretius, De rerum natura.

Here Lucretius, the Epicurean philosopher, shows the material necessity of the transfer of actual matter in hereditary process. He used it to prove the reality of unchangeable atoms but it may be used just as well to prove the existence of unalterable genes.

This short book with its innocently obscure title conceals the account of what is probably the greatest scientific discovery of our time, comparable with that of the splitting of the atom. And what an account it is! Because Jim Watson is a born writer and can write so amusingly that it is impossible to put the book down.

Here is the account of an epoch-making discovery by the discoverer himself and written fresh from the laboratory where it was made. Watson writes directly and uninhibitedly about work which he was doing himself or with his not so senior mentor Francis Crick. Both are highly individualist characters and their presence together at such a crucial moment in science is a further proof that the day of the lone scientist is not over. This work is the result of the unconsciously co-ordinated work of dozens of biologists and a group which will henceforth call themselves molecular biologists.

Watson worked fast at the double helix so he was unable to do what every young research worker is advised to do at the outset — read the literature. If he had done so he would have saved himself a lot of time and trouble. For the subject of the nucleic acids was a well studied one and their biological importance fully recognised. Since Caspersson's work, they had been taken to be the chief constituent of the chromosomes, the carriers of the genes of inheritance and it had long been known as well that they contained the nitrogenous basis, the pyrimidines and purines. The two best known varieties of nucleic acid, yeast nucleic acid and thymus nucleic acid were alike in this.

The next decisive clue to their structure was the discovery that they were high polymers and hence their solutions could be drawn into fine threads, a long chain of molecules like silk but containing the acid phosphoric groups. It was at this point that Astbury examined the fibres and found they were birefringent but negatively so, show- ing that the side groups were at right angles to the main fibre direction. From here the main interest had shifted back to the protein structure owing largely to Pauling's illuminating hypothesis of a non-integral helical and therefore only partially crystalline structure, itself an immensely liberating idea, which was to play a large part in the nucleic acid story.

I should say here that the distinction between the fully and partially crystalline structures was fully recognised in practice between Astbury and myself. I took the crystalline substances and he the amorphous or messy ones. At first it seemed that I must have the best of it but it was to prove otherwise. My name does not appear, and rightly, in the double helix story. Actually the distinction is a vital one. The picture of a helical structure contains far fewer spots than does that of a regular three-dimensional crystalline structure and thus far less detailed information on atomic positions, but it is easier to interpret roughly and therefore gives a good clue to the whole. No nucleic acid structure has been worked out to atomic scale though the general structure is well known. It may be paradoxal that the more information-carrying methods should be deemed the less useful to examine a really complex molecule but this is so as a matter of analytical strategy rather than accuracy.

A strategic mistake may be as bad as a factual error. So it turned out to be with me. Faithful to my gentleman's agreement with Astbury, I turned from the study of the amorphous nucleic acids to their crystalline components, the nucleosides. The easiest to prepare of these was cytidine. Very fortunately there came to my laboratory at that time a young crystallographer from Oslo in Norway, Sven Furberg, who had been working under Professor Hassel. He undertook the study of the structure of cytidine. He worked very quickly and well and the structure he found proved to have wide implications. In fact, had we realised it, it contained the key to the whole double helix story. But this key was never used because as far as we know neither Watson nor any of his friends either knew of it or realised its relevance.

There was a simple and entirely avoidable lack of communication here between men who knew each other and often worked in the same laboratories. I, myself, should have noticed it but I was too preoccupied with other things and so let the opportunity slip. This may stand as yet another example of a missed chance of an almost simultaneous discovery. I do not and will not make any claim to it for myself but I do think that, for historic justice, in the light of the importance of his work, Furberg's contribution has been grossly overlooked. By that time I had heard, in discussion with Crick, a nearly accurate account of the genetic code, so I fully recognised the importance of base order and I should have tried to work it into the X-ray picture of cytidine — another example of missed opportunity.

By this time, also, Rosalind Franklin — Rosy, in the book — had come to work in my laboratory. I had come to know and respect her and to admire her too, as a very intelligent and brave woman who was the first to recognise and to measure the phosphorus atoms in the helix, which proves to be the outer one, thus showing Pauling to be wrong and the helix to be a double one, though this inference is not drawn. Thus, all the elements of the structure of the full solution had been given though it remained to fit in the genetic evidence. This proved quite easy as the complementary self-replicating character of the double helix was almost self-evident. The clue to this was given by Chargaff who had observed that the sum of the pyrimidine bases was equal to that of the purine bases. Thus they could occur in natural paired bases, the pairs being held together by hydrogen bonds. Chargaff has never been given credit for this decisive discovery. Watson had played with the idea of a triple helix, which was absurd, and therefore missed the beauty of the two complementary strands. The mechanism of the transference was complicated but once that was achieved the rest was just filling in biological deduction though the biological consequences were to prove so far-ranging, offering the molecular key to the whole of reproduction.

All this is represented in *The Double Helix* in the form of a personal story in which the adventures of an American new boy in Cambridge society are fully and frankly described. College parties in rooms all come into it but despite the digressions the double helix is the hero of the piece and the naive pursuit of its structure the central theme. The book, therefore, becomes a kind of Moby Dick of the labs. In its pursuit and final capture we experience all the thrills without the dangers of the chase and all told without the conventional British self-effacing reticence; quite the contrary. Watson's self-revelations are often to the English mind positively embarrassing.

Here he paints a picture of science as it really is — an interplay of ambitions, rivalries and intrigues which can only be guessed at from the printed texts. But is this picture true? No and yes. Watson tells all he knows but not all there is to know as I have shown. If he had done so, the story would have been shorter but less dramatic. More justice could have been done to Pauling's many decisive contributions instead of his being treated merely as an unsuccessful rival — which he was not. In all, Watson's colours are too bright, too unshaded for a scientific work. Nevertheless, it is a book about the greatest scientific discovery of our age and fully deserves to be the bestseller it has become even if it is banned by Harvard and has incurred the wrath of the whole scientific community.

The picture of the making of a new science as well as the genuine account of how it was done contained in *The Double Helix* will be in itself a landmark in the story of science. It may serve to show how such a decisive breakthrough in human thought is not necessarily the work of an individual genius but only of a pack of bright and well-financed research workers following a good well-laid trail.

Appendix 4

James D. Watson's Foreword to The Road to Stockholm: Nobel Prizes, Science, and Scientists, *Oxford University Press, 2002.*

Foreword

The Nobel Prize is revered today almost as much as 50 years ago. Its several Nobel-bearing faculty enhanced greatly the spirit of the University of Chicago when I was one of its undergraduates. More important to me, however, was the 1946 awarding of the Nobel Prize in Physiology or Medicine to the 56-year-old American geneticist, Hermann J. Muller. His presence was the reason why I applied to Indiana University for fall 1947 entry to their graduate school. This Southern Indiana school was still in the intellectual backwater, and I would have thought myself down marketing if Muller had not just moved to Indiana. Though I had been told that Indiana also then possessed several young genetics hotshots, their names themselves would not have led to my going there.

As soon as I arrived in Bloomington, however, I discovered that Salvador Luria and Tracy Sonneborn were much more to my liking. They were much younger and did their research on the genetics of microorganisms, then genetics at its most exciting. In contrast, Muller's research still focused on *Drosophila*, the organism that he used for his Nobel Prize-winning 1926 demonstration that X-rays induce mutations. Twenty years later, work on the tiny fruit fly had a tired, almost arcane, feeling. In contrast, Muller's course on Advanced Genetics, which I took during my first term at Indiana, was an eye-opener. It made by itself my coming to Indiana worthwhile. No one else had lived through so many great moments of genetics, and I eagerly went to every one of his lectures knowing I would learn more how knowledge about heredity unfolded. In going to Indiana, I wanted to get close to the true essence of the gene. Over the next three years, I greatly benefited from the thoughts of an individual who had come to this objective some 35 years earlier.

Seeing Muller in action was equally important in letting me learn first hand that Nobel Prize bearers are not super humans. He was very much the seasoned academic, more at ease with the past than with how to move into the future. But I, having no past, had no choice but to gamble my future on a path that did not yet exist. To make my mark, I had to settle on a big objective and stick with it until either I or someone else got to the top. Here my Indiana experience gave me the factual knowledge I needed. I had gone there with no interest in DNA, but by the time I left to seek my fortune in Europe, I wanted to think about nothing else. Being then obsessive in no way guaranteed my way to the Nobel Pantheon. But it sure helped!!

James D. Watson
(19 November 2001)

Index

213